Praise for Bob Bancks

"With remarkable recall of day-to-day details punctuated by life-changing events, Bob Bancks brings back a way of life in an era when family farms were truly the work of whole families and also the foundation of rural communities. *The Adventures of Bobby, Iowa Farm Boy* gives readers a window into a past as close as living memory yet far away from the rural experience of today."

—Tom McKay, author of *West Fork* and *Another Life*

"Reading *The Adventures of Bobby, Iowa Farm Boy*, it was easy to imagine myself sitting in the porch rocking chair while listening to Bob tell his story. I identified with a lot of it, having grown up on a farm and having similar experiences. Bob will have your interest all the time, and you'll laugh at many of his tales, such as the time he and his sister decided to 'play house' in the basement while the adults were upstairs. You'll need to listen to Bob tell it to get the full humor of the situation. Needless to say, I highly recommend your listening to Bob tell his story."

—Alan Arkema, author of *The Letter*

"For those who didn't grow up in the Midwest, *The Adventures of Bobby, Iowa Farm Boy* is a superb introduction to what twentieth century life there was like. Bancks writes with the simplicity that belies the rural life he describes. His memoir takes the reader back to a less encumbered life, an honest look at the homes, lives, and industry of farming in Iowa. He describes a lifestyle devoid of frills, yet replete with plenty. It will give those unfamiliar with this life the comfort of being there in the moment. If you didn't grow up in rural Iowa, after reading this, you'll wish you had."

—Dan Moore, author of *The Last Voyage of the Marigold*

Other Titles by Pearl City Press

The Last Voyage of the Marigold by Dan Moore

Married, Living in Italy by Misty Urban

Books by Writers on the Avenue

Winter Holidays in the City of Pearls

Climbing the Hill of Life

From River to River

Everything Old is New Again: 30 Years of WOTA

Books by Bob Bancks

The Nightgown

Call Sara

The Fourth Generation

Iowa Exposed

There Are Bears

Oak Tree Tales

Edie

I Will Love You Forever

The Adventures of Bobby, Iowa Farm Boy

BOB BANCKS

Muscatine, Iowa

Copyright © 2022 by Bob Bancks
Cover design by Expressions by Em
Editing by Misty Urban
All rights reserved

Published in the United States by Pearl City Press
An imprint of Writers on the Avenue

All events are recalled to the best of the author's memory. Any errors and inconsistencies must be attributed to the passage of time and the unavoidable consideration that memory is an unreliable narrator.

For permission to reproduce selections from this book, contact Pearl City Press pearlcitypress@writersontheavenue.org

ISBN-13: 978-1-7369498-3-2

DEDICATION

I dedicate this memoir to my parents, Carl and Edna Belle Bancks. It was because of their guidance that I continued the Bancks Family tradition of farming the same land for 150 years.

To my grandpa, Charles Shepard, and uncles Vernon and James Shepard. They were my pseudo-fathers after Dad died.

To my elementary school teachers at Hazel Dell #3 for their devotion to teaching, despite the difficult times of the 1940s. They formed my education.

Lastly, to my neighbor, Wayne Kraft Sr. His continued guidance when I started farming was much appreciated.

Contents

Prologue	1
The Bancks Farm	3
Here I Come	6
Home Life	15
The Paulsens	20
The Second of My Nine Lives	23
Playtime	30
Wayne Kraft Arrives	37
My Chores	40
Battle Scars	53
My Pets	59
Hazel Dell #3	61
My Teachers	68
The Student's Day	77
Winter Fun	89
Allen	105
Scammers	109
Mary at the Bat	113

School Memories	115
Eighth Grade Exams	127
Sweetland Methodist Church	130
Miss Montpelier	145
The Cousins	149
Travels With Dad	179
Our Family Adventures	193
Modernizing the Farm	204
The October Sunday	217
After Dad	229
The Krafts	242
My 4-H Projects	263
Mom and Me	277
Ron	292
High School	297
Marriage and Beyond	301
About the Author	307

The Adventures of Bobby, Iowa Farm Boy

Prologue

ALTHOUGH I WAS NAMED Robert Charles, I was known most of my early days as Bobby. I was a blond-haired, hazel-eyed farm boy who was barefoot most of the summer. I had many adventures on the Bancks family farm. I didn't have computer games because there were no computers. Television existed, but not for rural folks. We had radios with backs full of glass tubes. I listened to fragile phonograph records and played them at a fast speed of 78 RPM.

My dad was a handsome six-foot-tall man with dark hair. He seldom frowned or was sad. He loved his family and his neighbors. He wore blue bib overalls and chambray shirts for work. His cap was striped denim. When he dressed up, he always wore a tie with a diamond stickpin. His favorite tie was green knit trimmed yellow. It was the tie he wore to meetings other than church. When he went to church, he wore a suit, white shirt, and a wide flowery necktie. In the winter he wore a felt fedora. In the summer he wore a straw fedora.

My mom was a small woman, five feet, four inches tall, and very thin during her early years. She had dark hair. She seldom wore slacks or pants until she was in her sixties. She wore dresses to town and house dresses at home. I never saw her without eyeglasses. She was devoted to her family, and she loved my dad.

My sister, Mary Ann, was four years older than I and five years ahead of me in school. I know we did things together, but because of the age gap, her interests were much different. I have few recollections of what she did in my early years. I know we never had a serious argument. It wasn't until we were both adults that we became close. Although she lived in California the last forty-some years, we maintained a close relationship. She died recently from cancer. I am the only one left to tell the Bancks family story of life in the 1940s and 50s.

My life was not without tragedies and failures. I learned how important every decision was to my family. I learned how to work at an early age. My chores although small were important. I saved Dad time by bringing in the cows for milking or saved Mom time by hunting the eggs and feeding the chickens. My dad was head of the household, but my mom was the disciplinarian. A swat on my fanny or time out sitting in the corner was my punishment. If Dad reprimanded me, which was seldom, I knew never to do that again. My dad didn't have time for discipline.

I learned early on that the animals and birds depended on Dad and me for their livelihood. They had to be cared for whether the weather was cold and wet or hot and dry. Even if you were not feeling well, you had to do your jobs. I could play all I needed, but work came first.

My dad had two philosophies. The first was "You should never think you are better than your neighbors." The second was, "There are seven days in a week; you work six of them." Many of those days were ten to fourteen hours long. He respected the men in town who worked forty hours a week, but he claimed successful people worked sixty hours or more. I have lived by these two axioms.

The Bancks Farm

IN 1866, JOHANN Hans Beenck and his wife, Catherine, left their native Germany for America. The newspapers in Schleswig, Germany, told of cheap, rich land in a new state named Iowa.

They landed in New York and traveled by train to Pittsburgh, Pennsylvania. There they boarded a packet boat and traveled down the Ohio River to the Mississippi River, then upriver to the little town of Buffalo, Iowa. Buffalo was located below a treacherous rapids on the Mississippi.

Catherine and Hans, as he was called, disembarked with the intention of catching the local train or stagecoach to LeClaire, Iowa, and continuing their journey upriver, but it was late summer and a poor time to travel, especially when Catherine discovered she was pregnant. Hans found a job working for Mr. Joseph Saur as a hired hand just north of town. He worked there for two years.

In 1868, Hans found a forty-acre farm for sale by Herman Karman in Muscatine County. It lay five miles west of Buffalo and five miles north of the village of Montpelier. Blue Grass, Iowa, was just three miles away to the east.

The forty acres had level ground and some pasture. It seemed like a good buy. So, instead of traveling on to central Iowa or beyond, in February of 1868 the Beecnks moved to Montpelier Township. They lived in a log cabin near a good spring and woods. During the summer, Emil

Robert was born. That fall, an adjacent forty acres was offered for sale by a Mr. Ehrecke. Hans purchased that forty. It was the beginning of several land purchases by the Bancks family. Today the farm encompasses 664 acres.

You may be wondering why it is the Bancks Farm and not the Beenck Farm. You can credit the name change to my grandfather Emil. During World War I, in which the U.S. fought from 1915 to 1918, the German population was not favored by some of the locals. Emil Beenck decided to change his name to a more American style of Bancks. It almost sounded the same if you put that upside-down 'v' or caret above the double 'E' in Beenck. It has been that way ever since.

One of my most cherished possessions is a document dated 1889 that declares Johannes Hans Beenck a citizen of the United States of America. He had to denounce any allegiance to the King of Prussia. I doubt Catherine ever became a naturalized citizen. She couldn't vote or hold any office, so why bother.

The farm buildings sat back from the county road about two blocks. It was an old German custom not to build buildings on ground that could be planted with crops. Another reason for the setback was that in 1868, you needed to be near water. Just north of the original log cabin was a spring. Great-grandpa Hans dug a well that provided water for livestock and the Bancks family for many years. The cabin was close to timber. Timber provided wood for the fireplace and logs for buildings and fences.

By 1870, the cabin had an extension built on the south side. It was constructed of sawed lumber, not logs, and it doubled the size of living quarters. I never lived in the log cabin, but I have a photo of my father and my aunt Georgianna sitting in front of the building when it was being used as a shed for cows.

The addition was converted into a granary when the new house was built in 1878. On the walls inside of the granary were paintings of a man's face and a steam locomotive. One of the Beecnks had an artistic flair and extra black paint.

Having your homestead away from the road had advantages and drawbacks. The advantages were less road dust, your neighbors weren't as nosy, and it was generally quiet. The disadvantages numbered few. Snow removal was always and still is a problem. If you were a kid coming home from school, once you reached the lane, you still had a distance to go. The mailbox was a long way down the road.

Today I cherish the long lane. The county road is still a gravel road with dust rolling after each truck or auto, but much less dust reaches the house this far away. The lane sometimes deters unwanted travelers, and seldom do people or vehicles arrive unnoticed.

I like being away from the prying eyes of neighbors. I can still watch the traffic until the corn grows too tall. I have a deal with my renter that requires him to only have corn on one side of the lane each year. You'd be surprised by how many times on hot summer evenings my wife, Jane, and I sit in our swing chairs wearing our pajamas. When family or friends visit, many of them comment on the quietness of our location. "You can hear the birds singing," is one of their comments.

The other comment comes in the evening when the sky is dark. "Gee, you can see the Milky Way out here."

Here I Come

MY MOTHER LIVED on a farm and was a schoolteacher. She taught at Hazel Dell School #3, which was located near the Bancks farm.

She stayed at Ellery Watts' place during the week. Mr. Watts would pick her up each Sunday night at an interurban station north of the school. On Friday evening he'd take Edna back to the station and she'd ride the interurban to her home. The interurban was a streetcar-like train that ran from Clinton, Iowa to Muscatine, Iowa. The rail track ran just north of her home near Muscatine.

On the warm dry Fridays in the fall and spring, she'd walk to the train station instead of having Mr. Watts drive her. Being a farm girl, she knew he was very busy during this time.

The short cut to the station, more than a mile away, went through the Bancks Farm. The only problem with the route was that she had to cross the creek in the Bancks' pasture.

One day, my father decided to walk my mother through the pasture and help her across the creek. Soon he was driving Miss Shepard, my mother, in his big Studebaker car to the station. They began dating and the rest is history, as they say. I found a written paper from Mom's bridal shower telling the story.

There was a time when Mom and Dad thought they would have no children at all. Infertility clinics didn't exist. They tried for children for six years before my sister Mary Ann was born in September of 1936. She was their little darling and they thought she would be their only child, but a surprise pregnancy occurred four years later.

Were they excited! The technology of determining whether it was a boy or girl didn't exist. They didn't care.

Mom was due in November. Dad was hand-picking ear corn at Fuzzy Palmer's, a neighbor who lived just north of our farm. As her due date drew closer, Mom hated for Dad to be too far away.

They devised a signal system. If she felt it was time to go to the hospital and Dad was picking corn at Palmer's, Mom would hang a white sheet on the outside clothesline which he could see from Fuzzy's. He would check every so often. Even Gladys, Fuzzy's wife, would check from the upstairs window.

Fortunately, he never had to hurry home. The morning of November 25, 1940, Dad was finishing morning chores and coming to the house when Mom decided it was time to go to the hospital. They dropped Mary Ann off at Grandma Shepard's on the way to Muscatine.

Dad drove Mom in our old Studebaker to Hershey Hospital in Muscatine. At 12:45 pm, Robert Charles Bancks was born to Carl and Edna Belle Bancks.

I was named after my paternal grandfather, Emil Robert, and my maternal grandfather, Charles James. My nickname would be Bobby or, later, just Bob. Dad wanted to keep names simple.

A few days later, they brought me home. I don't know the particulars, but I assume Mom returned to her duties as a farm wife. Now she had me to feed, Mary to care for, and Great Uncle Henrich, who lived with Mom and Dad.

It was a full household. When Mary was born and Mom was no longer available to help with the milking,

Dad hired a young man from the small village of Montpelier named Don Ellison. He would come early and milk cows, stay all day, eat dinner at noon, and maybe supper in the evening before returning to his home. I can't imagine how much food Mom had to prepare each day. She had Dad, Grandpa Emil, Uncle Hen, and Don to feed, plus taking care of Mary.

Don worked for Dad until the Japanese attacked Pearl Harbor, Hawaii, on December 7, 1941. Japan declared war on America. Germany and Italy soon joined the fray. America was at war in two theaters.

I was one year old. Soon all the young men in America were in the armed services. Don Ellison joined early and served in the Navy. Now Dad needed to find a replacement for Don. There were no other young men to hire since most were drafted and serving in the war effort.

Burt Paulsen arrived in the nick of time. I don't know where he came from; he just arrived with his wife, Vesta, and three children. He needed a home, and fate stepped in. A neighbor named George White was an older gentleman and became depressed because his son was drafted. Dad tried to help him with his forty-acre farm. One day while Dad was at George's farm, George told dad he wanted to sell his farm and move to Blue Grass.

Dad hurried home and got his checkbook. He wrote George out a check before he could change his mind. Now Dad had the needed house and buildings for Burt and his family. Burt stayed for the duration of the war. I will tell you a little later of my adventures with the Paulsen children.

With a great war going on, everyone pitched in, from four to ninety-four. In some ways, I was lucky. My dad was considered too old to be drafted and the nation needed farmers to grow food and fiber for the war effort. My dad helped the war effort by harvesting oats and soybeans for neighbors whose sons were fighting overseas.

He received an award from the extension service for combining three hundred acres of crops with his Allis-Chalmers combine. Today this would be nothing, but in 1944 this was quite an accomplishment with a small pull type combine cutting a five-foot swath. Dad spent many hours harvesting for others and would come home to do chores in the dark.

Everything was rationed. Gasoline, tires, sugar, and silk were the scarcest items. Fortunately, nature's natural sweetener, honey, was available to the Bancks family. Dad and Uncle Hen knew of several bee trees with bee colonies. They contacted a beekeeper to come with his smoker. The smoke calmed the bees in the tree. The tree was cut down and the honey retrieved. Of course, when the bees realized they'd been robbed, they were upset. Thank goodness, bees have a short memory. Soon they'd gather their queen bee and find another hollow tree. The honey was split between the beekeeper and the Bancks family. We all had some of nature's sweetener for a while.

By the end of the conflict, I was old enough to remember small things, like the Blue Grass Grocery Store delivering supplies that were not in the regular brown paper bag or cardboard box, but rather in a used sugar sack. Sugar was not purchased in five-pound bags; instead, the grocer dipped sugar out of a bin and weighed the white crystals on a scale. The used sack was how he received the sugar. Everything was in short supply.

Mom delivered eggs to the store in return for other supplies. With my Great Uncle Hen's help, we had a huge garden. Mom spent most of her days canning fruit and vegetables. Shortly after the war ended, Dad somehow bought enough copper tubing, a compressor, and insulation to build a deep freezer. He hired a man to build the unit, which was huge, over thirty cubic feet. We stored meat in one of the bins, vegetables and fruit in another, and the third was for the hired hand's food. As part of his salary, he received pork, beef, and chicken.

As with most farmers, we were almost self-sufficient. The only food items we lacked were sugar, flour, coffee, and salt. I found some ration books in Mom's things when she died. Each person in our family, including a very young me, had a ration book. Mine still had some stamps for tires and gasoline. No doubt I'd wished they were redeemable for toys.

Our great uncle Heinrich lived with us. He was Grandpa Bancks' bachelor brother. Everyone called him Hen. In today's world, he probably would be moved to an assisted living facility or even a retirement home, but in the 1940s, families cared for their aged relatives at home. He ate his meals with us and his room was on the first floor. No one ever questioned his right to live with us.

Uncle Hen tended the garden and lawn. His hibiscus flowers, which had a special spot in the garden, survived after his death for many years. His peonies are still living in our flower beds. All the fruit trees were his responsibility. He helped Mom with the canning, which was a huge job. Fruits and vegetables were stored for winter's use. Uncle Hen spent many days walking the pastures and fields searching for Native artifacts. His collection of arrowheads was extensive. During the winter, he'd take a sack of ear corn and walk through the snow to feed the quail and pheasant.

His job in the morning was to babysit Mary and me while Mom milked. He wasn't very good, but he could keep us from getting into trouble. He'd make sure we were eating breakfast before he lay on a sofa in our long kitchen/family room.

One day, like usual, he lay down and started snoring. Mary and I ignored the noise. But this day, his snoring stopped. When Mom returned from the barn, she tried to wake him. He didn't respond. She hurried to the barn and got Dad. I stood and watched as Dad tried to revive him. Finally, Dad relented and realized Uncle Hen had passed away in his sleep.

I was too young at three years old to realize what death was. My life changed much because of Uncle Hen death. Mom and Dad moved downstairs to the bedroom he had occupied. I moved to their bedroom and so I had my own room at the top of the stairs. Mary had to pass through my room to enter hers.

I don't remember his funeral. At three years old, I was rushed off to my maternal grandparents and stayed there. Funerals were for adults, not children.

Uncle Hen was the last of his six siblings to die. Three were born in Germany and three were born in Iowa. The Bancks family moved to the next generation.

When I was young, Mom contracted mastoiditis, an infection in her mastoid. The mastoid bone is located right behind the ear and the infection was very painful. She spent many days in Mercy Hospital in Davenport. Her condition was critical.

It was late spring, and school was dismissed. Mary and I were shipped to Grandma Shepard's. One day after church, Dad drove Mary and I to Davenport to see Mom. Children were not allowed beyond the lobby of the hospital, so we sat in the car and waved at Mom, sitting at the window of her room.

Mom spent several weeks in the hospital and many weeks recovering. Later she showed us the depression behind her ear where the infection was drained. We were fortunate Mom survived.

My sister and I had a good relationship even though she was four years and two months older than I was. Only once in my memory did we embarrass our mother.

Mom was attending the local Women's Farm Bureau meeting. Our neighbor Mrs. Noll didn't drive, so Mom offered her a ride. Mrs. Noll had five children. On the way home, Mrs. Noll discussed her problems with bringing up five children in a small house. Mom bragged about how well behaved her two offspring were.

The ruse probably would have worked, but for some reason Mom invited Mrs. Noll to our house. When she opened the kitchen door, I was chasing Mary around the kitchen table and screaming loudly.

Evidently, Mary had taken something of mine or teased me. The reason is long forgotten. All I remember is we were both reprimanded by Mom as soon as she returned from taking Mrs. Noll home. Reprimands meant a couple of swift swats on your fanny from Mom, and sometimes other punishment. I don't remember Mary's punishment, but I spent time on a chair facing the corner by the refrigerator. I was surprised to see cobwebs there.

Mary and I were often accomplices, too. One of our earliest escapades had to do with the rationing during the war. Mary and I were to be up and dressed every morning at seven-thirty. At that time we slept in the same room. We were awakened by a wind-up alarm clock.

If it was chilly, I'd dash for the warm air register behind the door in Mom and Dad's room next door. Mary Ann claimed her hot spot on the other side of the wall in our room. We dressed and scampered downstairs to the warm kitchen. We never ate breakfast in pajamas except for Christmas morning.

We'd set the table with the breakfast dishes and wait for our mother. Soon we heard Mom coming in the back door. She would kick off her four-buckle chore boots and hang her denim chore coat. She had been up since five to help Dad milk our twelve cows by hand.

"I'll get you kids your breakfast, then I'm going back to the barn and help Dad finish milking," she told us as she opened the kitchen door.

Mom would get the box of Cheerios out of the pantry and pour each of us a bowl. Mary got the milk, which was stored in a big copper pitcher in the refrigerator. We all drank raw milk in those days. Mom would just dip it out of the milk can down in the barn when we needed more. After drowning the cereal in milk, Mom dipped a teaspoon into the sugar bowl and spread each bowl with one small scoop.

"Mary Ann, you wash the dishes after breakfast. Bobby, you can dry the little pieces. I'll be back in less than an hour. It's cold today, so I'll drive Mary to school," Mom said as she disappeared out the back door.

I put two slices of bread in the toaster and pushed the handle down. Mary Ann ran to the kitchen window and watched Mom as she trudged to the barn. As soon as Mom was inside the barn, Mary Ann returned to the table, took her spoon, and scooped an extra portion of sugar for each of us.

"Now don't tell Mom or we'll both get paddled," she told me.

"I'll never tell," I answered with a smile.

This was the beginning of a long brother-sister relationship. Of course, we had some spats, which I seldom won, but we never told Mom about the sugar. I mean we never *ever* told our mother about our deed.

Home Life

MARY WAS IN SCHOOL from September to May. Even in the summertime, she would rather play girl things. She had an extensive collection of paper dolls.

The only time she played with me was during the winter. The nights were long, and her room was cold. We would play an extended Monopoly game on many evenings. We set up the game board on a small rug which we slid to the side of the room each night before bed. We would continue the game the next night.

In the summer, Mary Ann's interests were much different than mine and she decided there were other things more fun to do than playing with her younger brother. Like most kids, she had dolls; one was named Mary Jane and the other was Baby Jean. She'd play with her dolls up in her room. I even had a doll named Caroline. With all my girl cousins, I had to play what they chose if I wanted to be included.

On winter nights, Dad read the newspaper and Mom read or sewed. At times we listened to the radio programs of Bob Hope, Jack Benny, The Life of Riley, Fibber McGee and Molly, This is Your Life, and many others. Radio was great entertainment.

There were nights when Mary had homework or just wanted to read. Those evenings I had to devise my own entertainment. I'd play board games such as checkers, Monopoly, and Parcheesi by myself. Each time I would move around the board as a different person.

Of course, this way I always won. I could cheat and no one cared.

I had a collection of small plastic cars and trucks. Most were just the right size for small hands. These were either birthday gifts or toys I had purchased with my weekly allowance of ten cents. Mary took piano lessons from a teacher in Davenport, and while she was at her lesson, sometimes Mom would shop downtown Davenport. I seldom wanted to follow her, so she would drop me off at the north entrance of Kresge's Five and Ten Cent store. My orders were to stay inside the building until she returned.

I was six or seven years old, and my mother never felt I was in danger. I wasn't afraid of being left because I knew she would always come back and find me. Inside the rear door and down three or four steps was the toy department. I would spend the next thirty minutes ogling toys and dreaming of which car or truck I would purchase. Some only cost a dime, others a quarter. The dime ones had plastic axles and were not very durable. The more expensive cars had metal axles and withstood rougher play.

I knew not to ask for extra money. If I wanted a more expensive toy, I had to save my allowance until I had enough. There was a brand of plastic cars and trucks made by a company named Banner. Every month or so they introduced a new vehicle. The Banner vehicles were my pride and joy. These toy trucks were what I played with on the living room floor during the long winter nights. I would bring in small throw rugs and place books under them to make hills. Every night before bedtime I picked up every car and truck and carried them to my room. I was not allowed to leave toys lying around where someone might trip over them.

There were times when new models were not available or I didn't have the funds to purchase my selected truck, so I would go to the lunch counter and watch the automatic doughnut machine plop out fresh oil-fried doughnuts.

It was a fascinating machine. A nozzle squirted a blob of dough into hot cooking oil. A little conveyor eased the dough along until it was fried on one side, then a lever popped out of nowhere and flipped the dough over. A few minutes later, a fully cooked doughnut would drop on a wire tray to let the oil drip off. I hoped someday the girl behind the counter would give me a doughnut. It never happened. I'm sure if she ever did so, more youngsters would quickly show up.

I knew how to read time on the clock, so at the prescribed time I'd wait for Mom just inside door at Kresge's. She would park in the loading zone just long enough for me run out and jump into the car. Then we'd drive back to Mrs. Hershman's, the piano teacher, and pick up my sister.

I was never scared of being alone in the store. Maybe the store clerks were watching out for me, I don't know. There were always other children in the store. It was a different time. Right after the war people still respected each other. There was no taking advantage of a situation. People helped each other.

One Christmas, Mom and I went shopping at the big Petersen Department Store in downtown Davenport. The nearest parking lot was on the levee near the river about two blocks away. This day it was very cold and windy, and we had many packages. Mom sat me and all those packages on a bench just inside the west entrance door of Petersen's while she walked to the levee parking lot and got the car. I sat patiently until she arrived in the loading zone right outside the door. I did not leave the packages until she came inside to get me. That would be a very dangerous move in today's world. I may not have been kidnapped, but somebody would surely have taken advantage of a little boy watching Christmas gifts.

Because of my November birthday, I was almost six when I started grade school. Before I was deemed old enough to attend, there were some long days at home without Mary Ann.

Mondays were wash days. My time was spent in the basement with Mom. There was no automatic washer or dryer. Mom had a famous Maytag washer with a wringer mounted on the machine. She'd start with the whites, then do the colors, and finally tackle Dad's dirty overalls. In good weather she'd carry the clothes up the stairs and outside to the clothesline. In the winter she hung the clothes in the basement.

I'd help by pulling the clothes through the wringer and into the rinse tub. I never was allowed to start the clothes because my hand might become caught in the wringer. Sometimes I played in the basement while she washed. Dad fashioned a sandbox under the steps. The sand was not deep, but at least it didn't have cat poop in it like the one outside. I'm sure I scattered some sand around the basement floor. The floor was concrete but not polished, so the sand particles were not detectable.

Tuesdays were ironing days. I would help or attempt to help Mom. I had Mary's little play iron and play ironing board. Mom would let me iron Dad's farmer handkerchiefs. I guess he didn't care if they didn't have a crisp fold.

Wednesdays were baking days. I liked it when Mom decided to bake chocolate chip cookies. I got to lick the mixer beaters and maybe a small amount of cookie dough would be left in the bowl. There weren't any chocolate chips, just the dough, since Mom was stingy with the chocolate chips. They were rationed during the war, so very few chips made it into the cookies. Mom never forgot the rationing. Her cookies never had very many chips in them, even after the rationing was over. I learned to eat around the few chips in the cookie and save the morsels for the last couple of bites.

Mom's specialty was chocolate cake with a fudge frosting. Every cake was made from scratch, no box cake for her. Her frosting was delicious. Once spread on the cake it would harden and become like fudge candy. The best part was I got to lick or scrape out the saucepan.

She never had a recipe for the frosting; she'd just put the ingredients in a saucepan and boil them. If it didn't look right, she added a little of something. Once Mary asked Mom to write the recipe down, but she couldn't do it. All the ingredients were etched in her head. Mary tried several times to duplicate the frosting, but she never mastered the recipe. It was just one of those things unique to Mom.

The Paulsens

THE FARM DAD BOUGHT from Mr. Wright was primitive by today's standards. There was no indoor plumbing or running water. Water was pumped from a cistern. The living room had an oil-fueled heater. When Bert Paulsen replaced Don Ellison as our hired hand, he moved in with his family. It was a wonder his wife, Vesta, could raise three children in those conditions.

Two of their children were close to my age. Doreen was a year older, Dick was my age, and Danny was the baby. Vesta would babysit me at times when Mom had things to do in town. Let's say Vesta was around. She was busy with Danny and pregnant with another child. Doreen, Dick, and I were on our own most of the time. The three of us spent many days in the old shop making little boats to float in the small creek below the barn.

Dad furnished an old one-animal hog hut for Bert to store corncobs, which provided fuel for the cook stove. One winter day Dick and I decided to crawl into the shed and throw the corncobs from the small storage shed all over the lawn. When Dick's dad came home for lunch, he made us pick up all the corncobs and put them back. Dick got a spanking immediately, but because I was the boss's son, I was only scolded. When Mom came to get me, she was very upset. I lucked out on the spanking, but I did spend an hour sitting on my chair in the corner by the refrigerator.

During Bert Paulsen's stay, Dick was like most farm boys. He loved to ride tractors with his dad, just like I did.

One day my dad was cultivating corn to rid the ground between the corn rows of weeds and to aerate the soil. He was working in a field right next to Dick's house. Dick came running out, thinking it was his father. It didn't make any difference. My father knew he loved to ride.

This day Dad was cultivating with our B Farmall tractor. This was a small tractor used for mowing and raking hay, harrowing the fields, and cultivating. It could only accommodate a two-row cultivator. The driver sat on one side right over the row. The other side was an axle housing just the right size for a child's bottom. Dad stopped and Dick climbed aboard.

They made a couple of rounds. Dick evidently became drowsy watching the corn disappearing under his feet. All my dad remembered was Dick falling forward in front of the rear wheel. The rear tire ran right over Dick's head.

Of course, Dad stopped and jumped off. He expected to see a child lying motionless in the dirt. To his surprise, Dick jumped up and ran to the house, crying all the way.

Dad followed and met Vesta at the door. He explained what had happened. Vesta examined Dick. There were tire marks across the side if his head, but nothing else.

She said to my dad, "He looks okay to me. I'll give him a bath and put him to bed. Bert can look at him when he comes home."

It appeared nothing was wrong. The soft dirt and light weight of the left wheel never caused a problem. Dick's head was apparently very solid. It was one of those accidents which turned out not being tragic. A farm can be a dangerous place. Today we would hurry to an emergency room or urgent care to have Dick checked out. They would have x-rayed and examined Dick. There would be insurance papers to fill out and maybe an investigation by OSHA. All I remember was Dad saying Dick was okay. His head hurt some, but no other damage.

Bert and his family stayed until the spring of 1946. In March, Bert rented a farm east of Walcott. This is what was called the agricultural ladder. First you start out as a hired hand, then you save enough to buy some equipment to rent a farm, and finally you save enough to buy a farm. I missed Dick and Doreen, but soon I had another family to play with. I guess Dad planted the corn that spring with part-time help. Otherwise, I don't know how he would have accomplished the spring field work.

The Second of My Nine Lives

BEFORE BERT MOVED, Dad decided to cut down a large cottonwood tree behind the machine shed. He and Bert worked for days chopping with an axe and sawing with a two-man crosscut saw to bring the big tree down. Chainsaws were not available to the general public. Finally, the cottonwood came down with a mighty crash. It took Dad several days to cut up the tree and haul the pieces away.

One night Doreen and Dick came over to play while their father finished chores. We climbed on the fallen tree trunk. It was about four feet off the ground. The bark of a cottonwood is very thick and deep. When it was time for the Paulsens to go home, they got down from the tree easily. As I slid off, my foot caught in one of the grooves of the bark, and I fell.

When I hit the ground, it hurt. I didn't break any bones, just skinned my knee and elbow. I don't remember crying, but I was scared. When I went to bed that evening, I didn't feel well. Mom thought it was just the shock of falling. She put me to bed early.

In those days we had a small white bucket with a red rim we used for a pee pot. Mary and I used it upstairs because the bathroom was located in the basement behind the furnace. That night, as I peed in the pot, my urine was full of blood.

Not one to panic, my mom called our family doctor. He told her to come to his office in the morning and he would have a look.

The next morning, I was still passing blood. We went to Doctor Schroeder's office. From there he sent us directly to Mercy Hospital in Davenport. After a few tests and a calculated guess, it was decided my kidneys were deteriorating. I was a very sick boy.

I was put in a room with another boy. This lasted for a short while. My condition worsened. Dad hired a private nurse to care for me in the hospital, and I was moved to a private room.

Fortunately for me, a new pharmaceutical drug had just been discovered. The miracle drug was called penicillium. Doctor Schroeder hoped it would help whatever was wrong with my kidneys. Whether he knew for sure or was just guessing, I got a shot every three hours for three weeks. My butt looked like a pincushion.

Mom later had her own diagnosis. She thought maybe it was cold medicine she administered to me most of the winter. Most of the cough medicines contained codeine, an opioid. It was available in over-the-counter cough syrups.

Mom bought me a stuffed bunny. He was pale blue and had pink eyes. I named him Pinkeye. He was my consolation animal after the nurses gave me my shot. Many a time he rubbed my little sore behind. I still believe the nurses used blunt, extra-long needles.

The private nurse, a friend of our family, stayed with me for a week or ten days until I was off the critical list. From there on the regular nursing staff took care of me.

They were nice enough. It was the Catholic nuns who roamed the halls who were not as forgiving. They brought me books, but I was scolded if I crumpled a page or dropped the book on the floor.

The bed I was in was three feet off the floor. At night the sides were raised, so I was in a little metal cage. Mom brought me a thick jigsaw puzzle. I put it together so many times I could assemble it upside-down. One day I accidently dropped a piece on the floor. I pushed the little button for help. A few minutes later a nun came rustling into my room.

She asked, "What do you need?"

"I dropped a piece from my puzzle, and I cannot reach it," I replied.

She gave me a stern look, picked up the piece, and said, "Next time you be more careful. I haven't got time for picking up toys. I'm busy."

I was only six, so I was scared. I never pushed my button again for that reason. I would wait until a regular nurse arrived for my shot.

There was one nun I really liked. She wore white while the others wore black. She was old and wrinkled, but so kind and sweet. She read stories to me and told me about the next meal. The other nuns called her Mother Superior.

One requirement of my stay was to drink a pitcher of water every eight hours. Early in the morning, a nurse brought in a pitcher and a glass with a glass straw in it. By afternoon, I sucked down my first jug only to have it replaced by another. If I was behind in my drinking, the nurses visited every few minutes to make sure I was getting my fill. Most days I accomplished my task.

The hospital food was something else. Most food was okay. Every few days for dinner, I could have Iowana ice cream, which was a green ice cream in a cup. Iowana ice cream was the premier ice cream company in Davenport. It wasn't my favorite flavor, but it was better than the pudding. Some days Mom would stop at a drug store a block away and buy some vanilla or chocolate ice cream for me. The store would hand scoop it into a pint container. That was a real treat.

The worst meal was when the kitchen served Cream O'Wheat for breakfast. I could not stomach that goo. I'd rather eat paste. To this day I cannot eat that ugly cereal.

The best breakfast was a little box of Cheerios. The nurse would open it and let me pour the contents into a bowl. She'd pour the milk, which was never enough. I'd let the little Os soak a bit before I ate them. I still like Cheerios.

My other treat was an extra chocolate chip cookie. My great-aunt Ivy worked in the hospital kitchen. When she baked cookies, she sneaked one or two in her pocket and hand delivered them to me before she went home.

Slowly, I improved. My mother visited every day. Dad visited on the weekends. When I needed a blood transfusion, he was my match. He came to the hospital early and gave his blood, then stayed with me as I received the precious fluid.

I remember him sitting by my bed and reading several books. It was the first time Dad ever babysat me. I never realized it, but my condition was critical. Without his blood I may not have survived.

Across the hall from my little room was a room enclosed by glass. It was a semi-isolation room. Parents would come and look through the window at their sick children. One evening a sailor dressed in his Navy whites stood at the window. Dad was visiting that day. Although the war was over, the military was still very active. Dad left my room to talk to the young man.

"What's the problem, sailor?" he asked.

"My little girl was severely scalded by a pan of boiling water she pulled off the stove," he replied. "I've got two weeks leave to come home and be with her. It took me three days to get here."

"Will she recover?" Dad asked.

"Oh, I think so, but it will be a slow process. The doctor wants to send her to Iowa City. They have a burn unit there. The unit is full right now, so we will have to wait for several days before we can move her."

Dad found out who the man was and where he lived. He told him he would contact my private nurse to see if she would help. Of course, she said she would. I don't know if Dad paid for her services or not, though I imagine she didn't charge the sailor's family. The little girl was there for several more days before she was transferred. The sailor came to my room. In his hand he carried a model ship.

He said, "This a model of a Navy destroyer, just like the one I am on. It was painted in camouflaged colors of blue, pink, green, and brown so enemy planes couldn't spot it. We chased Japanese submarines and tried to sink them. Sometimes, during big storms at sea called typhoons, the waves would go completely over the top of the ship. I want you to have this model of my ship because your father helped me and my little girl."

I still have that model, though it has had some bruises. As I grew older, I learned to cherish it more. I never knew the name of that sailor, but I'll never forget the gift he gave me. It was a symbol of gratitude by a sailor who needed someone to help. My dad could have sat and watched, but he intervened and supplied the help.

One evening, two teenaged men walked into my room. They were Wayne Plett and Wayne Bangert, who were good buddies and good friends of our family. Wayne Bangert was my cousin, although much older. I don't know if my Aunt Georgie or Wayne Plett's mom encouraged them to visit me, but it was memorable. They read me stories and helped me color in my coloring book.

Wayne Plett was a cut-up. He had a deep voice and a funny laugh. He found a drawing of Bugs Bunny and started to fill in the picture, only he added his personal touch.

He said, "I believe I'll make old Bugs an Indian. I will put a feather between his ears."

So he drew a feather with a headband. They didn't leave until the intercom announced visiting hours were over. I'm sure they spent the rest of their evening cruising the streets of Davenport.

I finally got to come home after four weeks in the hospital. Doctor Schroeder felt I could finish my convalescing at home. The weather was warming, and spring was just around the corner.

Dad and Bert were rendering lard from butchering several hogs when Mom and I returned home. We butchered our own hogs in late February. Rendering the fat into lard was one of the last processes. Mom pulled the car up near the garage because the rendering kettle was set smack in front of the garage.

I was so glad to see my dad. I probably could have walked into the house, but Dad carried me. Maybe I didn't have any shoes on; I don't recall.

I spent the next couple weeks in my parents' bed downstairs. They slept upstairs in my bed. It was easier for Mom to take care of me downstairs. Mom only let me out of bed to go to the bathroom. It took a week or so before I could play on the floor next to the bed.

Soon I could wander around the house, but not outside. I was thoroughly bored.

While I was confined to the bed, Mom brought in my meals. I had a small table next to the bed. One evening, before Dad came in from chores, she fed me early.

I finished my meal and waited for Mom to come. She was busy in the kitchen and was delayed. There was a table lamp next to my bed and I wondered, what happens if you pour water on a hot light bulb when it is still lit?

I picked up my almost empty glass and poured the liquid on the burning bulb.

Zap! It burned out.

Son of a gun! I was lucky it didn't explode.

When Mom returned, she asked, "Did that light bulb just burn out? I thought it was burning when I was in here last."

"I don't know, but it just went out," I replied, trying not to look guilty.

She said, "I guess it just was its time to burn out."

I wasn't going to confess that I was experimenting with my water.

I was sure glad to be able to get outside. I didn't like being penned up inside the house. By mid-May I was back to my old self, Mary was out of school, there were eggs to hunt, chickens to feed and pastures to explore.

When you are young, you soon forget your hurts. This episode would be the first loss of one my nine lives. There would be more trials coming later. I still drink plenty of water because of my kidney ordeal when I was six.

Playtime

MY EVERYDAY WEAR in the 1940s summers was shirt, bib overalls, or little boy's pants. In the winter, I wore leather shoes and warm cotton socks. Tennis shoes were not available for the common person. In winter I wore wool coats, snow pants or double overalls, flannel shirts, and wool mittens which soon became wet and cold.

I wore four-buckle rubber boots which in time became two-buckle rubber boots because the bottom buckles got torn off. If my boot became torn from climbing over a fence or a sledding accident, Dad patched them with a tire patch. We were not wealthy, so patches on boots and trousers were accepted.

The worst item Mom made me wear was cotton leggings. It wasn't so bad when I was home, but she made me wear them my first year of school. They were not like the stretchy elastic kind worn today. They were tan cotton hose held up by a garter belt and snaps.

They may have worked for little girls, but for active boys, these leggings were a disaster. The snaps would work undone. The cotton hose would sag. It was impossible for me to re-hook them without dropping my pants. At times I was really embarrassed. None of the other boys wore them. My sister finally came to my rescue and told Mom I was old enough that I didn't need to wear those ugly things.

In the summer, I was outside most of the day. My attire was minimal. I was barefoot except when we went to town or church.

The soles of my feet became tough as leather. I could walk across gravel roads with ease. The foot wash was a nightly ritual. My little feet would be filthy. Most days I never was allowed in the house until I scrubbed them first. Mom would stop me at the back door and hand me a small bucket known as a foot tub. Many times, my sister was my foot scrubber. After scrubbing the dirt and grass stains, she'd use an old towel to dry them.

This kept my feet clean for a little while, but remember this is summer, and returning outside after supper was not unusual. Before bed I just wiped my feet on the entrance rug and went to bed. I probably had dirt in my bed. I took a bath on a weekly basis unless I was dirty; then I was asked to shower in the basement.

Mostly, I played alone. I made up stories and situations. Even at four and five, I developed a vivid imagination. Early years confined me to the yard and garden for play. The yard was large, shaded by an oak tree and several large silver maple trees.

I had two rope swings. One we named "the little swing," which was suspended from a large pipe spanning between the oak and one maple. It was close to the house and shaded. The other swing, dubbed "the big swing," hung from a limb on a cottonwood tree west of the house lawn. Mary used it mostly because I was afraid of the distance the swing traveled. At the apex of the arc forward you could be fifteen feet above the ground because the tree was on a hill, and as the swing traveled, the ground below sloped away.

I don't recall anyone ever getting injured. I'd hold on tight when Mary pushed me. She'd push me high enough she would run right under me as I swung. At the top of the swing's arc, you could see Ira Dipple's house and barn on the other side of the cornfield.

One day a bolt of lightning struck the cottonwood and split it down the middle. The big swing was no more. We lost one of the greatest swings in the country. The little swing lasted until we built the new house. Both trees were in the way and had to be cut down.

My summers were as full as a young boy could want. I had my toy tractors, wagons, plows, and disks. I didn't have a sand box because, first, clean sand was difficult to obtain, and second, the barn cats loved to use the sand as their litter box. I played in the garden dirt between the rows of vegetables.

Mom raised tomatoes, peas, beans, carrots, and onions. The garden had a large asparagus bed, which was a good place to have play farm fields, and it was shady. The best area was between the rows of Swiss chard. This plant had stalks about eighteen inches tall. It was better than the asparagus because the soil was worked and soft. My toy field equipment worked well. I had to be careful not to knock down any of the chard.

I valued my toys. Every afternoon when it was time to quit, I gathered my toys and hauled them to the wash house, a little building just outside the house. At this time, the building was used for storage, not washing.

I never left my toys outside overnight, not because someone would steal them, but because the night moisture might harm my precious friends.

I had another unusual place to play farm. It was under the fruit trees in the chicken yard. Our chickens were not confined to a building in the summer but had an outside dirt yard containing peach, apple, and plum trees.

The hens rested under the shade of the trees. Now chickens, like other birds, like to dust themselves. In nature, birds dust themselves to rid their feathers and bodies of insects. Domesticated hens still follow the old ways. They made depressions or wallows in the dirt. Because of their scratching, the soil became very fine, almost like sand. I liked using their dust bowls for my roads and fields. The fine dirt made for good playing farmer, and it was always shady under the trees.

Occasionally I had to clear out droppings by scooping them up and tossing the gooey substance away. I'd get some of the droppings on my hands, but I'd just wipe my hands on my pants to clean them and continue playing. Even in the chicken yard, I never left my toys outside overnight. I was innovative at times and would cover the toys with a bucket or two. It saved a lot of carrying.

My summers were long, and I had no close playmates. Dad was busy farming. Mom had her housework and gardening. Mary Ann wasn't a cowboy fan. Heck, she didn't even like to play catch. It wasn't what girls did.

Frankly, I don't remember what she did to occupy her time. I guess she played with her dolls. Paper dolls were popular. They were cutouts of current movie stars like Shirley Temple and Margret O'Brien. They came in a book and Mary would cut out the figures and all their different outfits. She would create stories for her characters.

Our play helped us develop vivid imaginations. Mary became an excellent elementary school teacher. She loved to read stories to children. I guess I did the same with my toy trucks and tractors. I had fields and roads to mimic Dad's. My play in the pasture was developed by hearing stories and going to movies.

When I was seven or eight, my world expanded to include the pastures north of the buildings. Mom figured I would not get lost because I drove the cows home most evenings by myself. I knew to avoid the bull although we never had a mean one. I also knew not to try anything stupid.

The pasture nearest to the house had a small creek which ran water until July. If the summer was wet, it ran water all year. The pasture also contained a most exciting cottonwood tree. The tree had many roots exposed. I could climb over and under the curvy extensions. Sometimes I carried a few of my small toy cars down there and used the roots for roads. Some of the roots were dead and hollow. They made great tunnels.

One tragic day, I drove one of my cars inside the decayed root. It was a pretend garage. Later I added another car, and another. When I became tired of playing on the roots, I started to retrieve my parked cars. The first two were easily extracted, but the last car, my first in, had been pushed further back than I could reach. I tried to reach in the hole with a stick, but to no avail.

I gave up and went crying to the house. Mom thought I was hurt, but when I told her my sad story, she said she didn't have time to help me. Mary Ann volunteered to help extract the wayward car. We carried a garden hoe and a piece of baling wire to the mass of roots.

The hoe was too large. The wire could reach the car, but we couldn't find a hole to hook onto to drag the toy out. Mary sent me back for a flashlight. Now we could see the car, but the hollow root turned down. She worked and worked for several minutes but instead of coming forward, the car kept dropping further into the abyss. If

Dad had been available, he might have chopped the old root in two, but we knew he was too busy for such a small job. Mary and I gave up. I never recovered the car. But I learned a valuable lesson: Never play with your prized cars on the cottonwood roots where they might become part of the tree.

The pasture furthest away was the summer pasture. It was full of play possibilities. The east side contained a grove of oak trees growing in a steep ravine. The west side had an eroded area with yellow soil exposed. The soil was so poor that very little would grow there. The cuts and draws were great places to shoot the bad guys or play army. The north side contained a creek large enough to wade in. It was seldom dry, although in times of extreme drought it ceased to flow. Before the creek disappeared into our neighbor's pasture, there was a spring. The bank was always wet and black. Dad said it was a peat bog, sort of like pre-coal.

When the creek was flowing, minnows and frogs lived there. Water bugs scooted across the surface like ice skaters on a pond in winter. Mostly it was a mud bottom, but there were areas of sand. They were the places to wade and not get your feet and pantlegs dirty.

The days I went to the pasture to play my imaginary games, Mom packed me an empty Karo syrup pail with a peanut butter and jelly sandwich, a couple cookies, and carrot stick. My water was a pint fruit jar. Off I would go, playing all afternoon by myself. I guess Mom never worried about me getting hurt. It was a farm boy's life. You did your chores, you dried the dishes without dropping them, and you were careful with your toys. You could be trusted.

If the hickory nuts were falling, their shells made nice little boats. I'd float them down the creek. In warm weather, I could follow them by wading behind. On cooler days, a stick was the tool to move the shells along which floated out of the current.

Rainy days were especially long. If it wasn't too warm upstairs in my bedroom at the top of the stairs, I would fashion farm buildings out of cardboard boxes. My fields were the different squares and lines on the carpet. At mealtime, Mom would call me to set the table, which was one of my chores. If it was noon, playtime was just suspended. If it was suppertime, I would return the next day, but I made sure all the equipment was stored in its proper cardboard shed.

In the winter, my room was cooler because it was a good distance from the furnace, so I had to wear extra clothes to keep warm. Playing in my room was better than downstairs because I didn't need to carry my toys down and back.

Wayne Kraft Arrives

IT WAS LATE JUNE or early July, 1946. Dad was repairing our Allis-Chalmers pull type combine. It was early evening. I was small enough to crawl inside the machine to hold bolts while Dad attached nuts on the outside. We were working under the yard light next to the garage.

A 1937 Ford drove up the lane and stopped next to the combine. I heard Dad talking to someone. I stuck my head out of the straw shaker compartment of the combine. There was a tall skinny man visiting with Dad.

Dad said to me, "We're quitting for tonight. I need to talk to this man. Let me help you out."

Once out and on the ground, I looked over at the man's car. There was a lady and a child inside.

"Bobby, you go to the house and tell Mom we'll be there in a jiffy."

I dashed inside to deliver the orders. In a few minutes, Dad walked in with this tall man with a loud voice, his wife, and a little girl. They sat on the sofa and talked.

Mom said, "Bobby, get some of your toys for the little girl."

I did as I was told, but what do you find for a two-year-old to play with? I brought some cars down from my room. Where was my sister when I needed her? Dad and this man visited for about an hour. The girl found little joy playing with my cars and I was worried she'd break one of them.

Finally, Mom took the lady into what we called our living room. The little girl went with her mother. I was relieved. Later Mom found some cookies and iced tea to serve the couple before they departed.

After the couple left, Dad said, "Wayne works for Walt Egel and Walt is retiring. He seems like a good man."

Mom asked, "When is his wife due?"

"Late September. They should be settled in by then."

I asked, "Due for what?"

Mom gave the look that said *it is not your problem*, but she answered, "The lady is going to have a baby in September."

The next week Wayne and Valerie Kraft moved into the little house where Bert Paulsen had lived. Wayne and Dad hit it off right away. Wayne came from a large family who lived near Lena, Illinois. His wife was Valeria Best, whose family lived in Fulton Township. Their little girl's name was Gloria.

Wayne had only worked two weeks when he had an appendicitis attack. He had an appendectomy and was off for several days. I remember Dad was fixing something in the basement when Wayne returned. It was only about a week after his operation. He started carrying some bags to the basement, and Dad had to stop him. He didn't want any more complications.

Wayne was one of hardest working men Dad ever hired. He could carry two fifty-pound bags of feed instead of one. He could throw bales halfway across the barn. Wayne's best attribute was he could be trusted. If something broke down, Wayne could fix it. If Dad needed a fence built and he was busy, Wayne could build it. Wayne had a loud voice. My dad had a voice that would carry a long way. It was said the two men could carry on a conversation while standing half a mile apart.

Because of Wayne, our family had the opportunity to take two three-week vacations while he worked for us. One trip was to Yellowstone National Park, Glacier National Park, and Banff, Canada. The other three-week vacation was to Niagara Falls, Gettysburg, and Washington, D.C.

Wayne proved to be a great asset. He and Dad complemented each other. Wayne could do just about anything.

In 1948 a neighboring farm, which had been part of an estate, was for sale. Dad thought a lot of Wayne and his abilities. He bought the farm and financed Wayne for ownership. Because of Dad, Wayne skipped the renting rung of the agriculture ladder.

The first year, Wayne started with an ancient F-20 tractor and manure spreader. Dad sold him some milk cows. Wayne borrowed Dad's equipment to plant his corn. My grandfather was a Pioneer seed corn dealer. He sold Wayne seed corn and told Wayne he didn't have to pay for it until fall when the crop was harvested.

Dad had two other hired men after Wayne. He and Wayne still baled hay and harvested corn together. They remained great friends. Wayne was the first to help when Dad fell from the ladder. When he asked if he could farm our place after Dad died, Mom couldn't turn him down. It was the beginning of a long relationship with the Kraft family.

My Chores

BY THE TIME I was five, I was helping do chores. My duties were hunting for eggs, feeding the chickens, and feeding the bucket calves. We were a dairy farm at that time.

The chicken chores were the best. We had around one hundred and fifty hens. When the hens were laying, we'd have seventy to ninety eggs a day. Mom hunted them in the morning, Mary and I at night.

When Mary was practicing the piano, I did the chicken chores, which I enjoyed. Chickens are small animals compared to pigs or cows. They don't hurt your toes when they step on them. Sure, they peck you occasionally, but otherwise they are soft and lightweight.

I fed them chicken mash in little feeders. Mash was a mixture of ground corn, oats, and soybean meal. Because we didn't have good mixing equipment, Dad would take a load of whole corn and oats to the Blue Grass Mill three miles away. They added the soybean protein and other ingredients. When the mash was ready, they bagged it in burlap bags and loaded it into the wagon for Dad to take home. This process took about a day.

We stored the feed in the garage or in the corncrib if there was space. A three-gallon pail was usually enough to feed the chickens for a day. In the summer the feeders were placed outside. Once inside the chicken yard, I would call, "Here chick, chick." The chickens would come running and scratch my bare feet with their toenails in their hurry to eat.

In the wintertime, we mixed the feed with hot water and stirred it into a soupy mash. The hens loved this mess, especially when the weather was bitter cold. After Dad died, we purchased a prepared chicken feed which was made into pellets, then crumbled. It was much easier to feed and far less dusty.

A promotion by the Ful-O-Pep feed company was to bag the feed into cotton print sacks. It took about a yard of fabric to make a bag. Farm women, my Mom included, would sew girls' dresses and skirts from the material, but mostly it was used for napkins, tablecloths, curtains, and aprons. It was a gimmick which sold a lot of chicken feed for a while.

The chicken house was not heated, nor did it have running water. Every morning and night we poured a five-gallon pail of water into a waterer designed just for chickens. It was a cylinder over a cylinder. There was a small drinking trough around the outside. As the chickens drank, the water would expose a little hole. The hole would let air inside and water out. If the water covered the vent hole, the water stayed inside the supply tank.

Lifting a five-gallon pail of water was difficult for a six-year-old boy. My mom could do it with ease. I would pour half the water in a smaller pail, dump it, and repeat the process. I had to work fast because the waterer without its cover let the water run out freely. In later years, we installed an automatic heated waterer.

Most eggs were laid in square boxlike nests which lined two walls at the west end of the chicken house. One set of nests was built so the rear door dropped down and the eggs were exposed for pickup. Most hens, if they were still on the nest, wouldn't mind me sticking my hand under their feathery rear ends to recover the eggs.

Occasionally, a hen would get her hormones in a huff and refuse to let me get the eggs underneath. Mother Nature told her it was time to sit on the eggs and hatch them. Unfortunately for the hens, we had no roosters; therefore, the eggs were unfertilized.

The hen would peck at my hand. To retrieve the eggs, I had to become aggressive. I would reach in and grab the hen's neck so she couldn't peck me. With my other hand, I would find the eggs. Sometimes, just for fun, I'd pull the hen out and toss her across the room. She'd go squawking and flying away, telling me how rude I was. I never hurt them, but I cured their stubbornness.

Not all hens laid their eggs in the nests provided; some dumb hens laid under the roost. I had to crawl under because I was the smallest person. Crawling under the roost to retrieve the eggs meant sometimes getting chicken droppings between your toes.

Now, chicken poop between your toes is ugly. It feels ugly, it looks ugly, and it is ugly to remove. The gray and white droppings stick like glue. You can't wipe this glue off in the grass. You might be able clean some from between your toes with a stick, but water and a brush are needed to rid your toes of the goo. Yes, I could wear shoes when doing chores, but the goo didn't happen very often. I had educated toes, and this was summertime.

I always liked chickens. Mom had a system of buying Rhode Island Reds one year and White Rocks the following year. Each spring, Mom would order three hundred chicks. The odds were that half would be hens. The rest were roosters. We housed the little hatchlings in a small building called a brooder house. They came as fuzzy little yellow chicks.

Mom made a point of waiting until I came home from school to unload the new birds. She knew I liked to lift each fuzzy chick, dip its beak in the waterer, and place it under the heated brooder. In a few days, the chicks began to sprout feathers.

As soon as the birds were fully feathered, we let them out into a small pen in front of the building. As they grew older and took up more space, we let the chickens run free. Eventually, in late September or early October, the young hens started to produce eggs. They were small at first and all over the farmstead. It was a challenge to hunt the eggs.

When the roosters weighed five pounds or more, they were butchered. Mom could do five or six cockerels a day. When Dad was around, he'd help kill the birds and pluck their feathers. Later, as I became stronger, I did that job.

The pullets or young hens we transferred to the laying house. This was done in the evening. The pullets would almost always return to the brooder house at night. As soon as it was dark, we'd shut the door, start grabbing the birds by their legs, and stuff them into a chicken crate. We then hauled the crate to the laying house and released the pullets. This would be their new home. In time, they learned to lay their eggs in the nests and became our reliable source of eggs.

Because each year Mom had hens of a different color, we could easily tell how old the hens were. We seldom kept hens beyond three years. The spent hens were butchered and used for chicken recipes such as escalloped chicken, hot chicken sandwiches, and chicken salad.

We culled the older hens with a quick process. First, you check the hen for color in her beak and on her legs. If they were almost white instead of yellow, it meant they were still laying. Second was the three-finger test. You place your fingers in the area where the eggs emerge. If you could easily place three fingers there, she was still laying eggs. Ultimately, hens older than three years were culled anyway. Their production would be down soon and there wasn't enough room for the old hens and the new pullets.

The chickens meant extra money for Mom. Every week we delivered one or two crates of eggs to a grocery store in Davenport. The stores were not regulated as they are now. Most could accept eggs direct from the farm and sell them in the store. A crate would hold thirty dozen eggs.

In the early fall, we'd have two crates of eggs per week. We would place them in the trunk of the car and drive to town. Mom would receive a credit slip to present to the checkout clerk. We raised chickens for many years until I went to college. After that it was too much work for Mom by herself, so the chickens were sold.

An old neighbor taught me how to hypnotize a chicken. I still can lay a bird on the ground, make the hex sign, and let her go. The chicken will lie there if someone doesn't make a quick movement.

I loved working with chickens and raised them for a 4-H project. Turns out, I was very good at raising chickens. There was a special show at the Swift Plant in Muscatine. I won the show. My prize was a tour of commercial poultry houses in Iowa. I was only thirteen and the tour was the same time as the West Liberty Fair, so Mom and I decided I shouldn't go. Another 4-Her took my place.

Another one of my chores was bringing the dairy cows home. Dairy cows are big docile creatures and don't need a lot of driving. Mostly they know to come home at milking time. Occasionally they got their clocks messed up and they would be at the far end of the pasture. I'd have to trudge after them and holler for them in my little boy voice. "Come, bossy, come bossy."

The cows would raise their ears and start their long trek to the barn. If you know anything about cows, you'll know that when a cow gets up from a lounging position, she immediately raises her tail and relieves herself. I was a master of running at full tilt in bare feet and never stepping in a fresh cow pie.

If you have ever stepped in a fresh cow pie barefooted, you'll find they are gritty and full of grass splinters. It doesn't hurt, it just feels funny, and is difficult to wash off, especially between your toes. Mom had a special brush she purchased from the Fuller Brush man. It was just right for sneaking in between dirty toes and cleaning out manure.

Once the cows were tied in their stanchions and eating, it was my job to take a barn broom and sweep the crumbs closer to the cow's head. The bad thing about this chore was sometimes a cow would take a drink from her drinking cup located beside her stanchion. She'd raise her head and throw slobbers all over. I didn't mind it on my pants, but it was slimy, icky stuff on my skin.

My dad was a medium-sized dairyman for the 1940s. He milked 20 to 25 cows twice a day. He carried the milk into the milk house and poured it into a vessel called a strainer. The strainer had a cloth disc which filtered out any dirt and other debris.

The strainer sat on the top of a ten-gallon milk can. Once the can was full, Dad or whoever was milking would lift it into the milk cooler, which was a tank of water cooled to forty-five degrees. Our cooler had space for ten milk cans or one hundred gallons of milk.

At the end of each milking there was always a little milk in the strainer which was too slow to filter through. Dad would carry the strainer back to the barn and call the barn cats. As soon as they heard the call, "Kitty, kitty," they'd bound in from everywhere. They'd drop out of the hayloft, sneak in through open doors or windows, and crowd around the pan. Dad poured the milk into the pan. The cats didn't mind if they got some on their heads. They had plenty of time later to clean themselves.

Farm boys are always busy. One day I spent the whole day relaying the orders from the hay rack to the young guy driving the tractor on the pull-up. The pull-up was a small tractor hooked to the hay rope to pull bales into the haymow. This was a very windy day, and the kid driving was having a difficult time hearing the man sticking the bale fork. I sat next to the asparagus bed and yelled, "Go ahead!" or "Stop!" I really felt important that day. Dad gave me a dollar for my work.

Baling hay required a big crew. Dad owned the baler, so he bartered help for baling. We would start at Lester Bohnsack's, then go to Ira Dipple's, then Wayne Kraft's, and finally do our place. Lester, Ira, and Wayne provided the needed extra help at everyone's place.

One day we must have had a bunch of hay to make. Dad had some extra teens to help. Don Plett was one of the extras. There was one person hauling the loads in, one unloading, two loading on racks in the field, two working in the haymow, one kid driving pull-up, and one man driving the tractor pulling the baler. I was bored because I was told to stay in the yard and out of the way.

At 3:30 Mom had me help carry out the lunch she and Mary had assembled. It included several sandwiches, cake, cookies, and plenty of lemonade or iced tea. The men working at the barn rode out on the last hayrack. It was great fun feeding the men lunch. There was always a lot of kidding and joking between them.

After lunch, I begged Dad to let me ride the rack with the men. I promised not to get in the way. I told him I would climb up the bales as soon as the loaders had some stacked.

Don Plett was one of the guys loading. He kidded me and tickled me. I liked Don. Dad relented and asked Don to watch me. I don't think Mom was so keen on the idea.

Dad decided to pull an empty rack behind the one they were filling because this way the racks could be quickly changed. Because of lunch, the men at the barn were slower getting back with an empty rack, and we were at the far end of the field, so we switched the racks and hooked the full rack behind the empty.

Since I was already on the loaded rack, I chose to stay there. The field had some hills, but not steep ones. I sat near the back of the full rear rack. Time moved slowly. I became sleepy. I tried my best to stay awake.

I remember grabbing for the twine on the top bale as I tumbled off the rear of the rack. It was about ten feet to the very hard ground.

Next thing I remember is waking up on the couch in the back porch. Mom had a damp wash rag on my forehead. My back hurt.

Don Plett came in to see how I was doing. He told me what happened.

"I looked over the top to see how you were and you were not there. I jumped from the rack and found you lying on the ground. You were out cold. John was coming so I picked you up, switched wagons, and drove home with you on my lap."

His next question was, "How's your back?"

Mom helped me sit up and turn around. My back was a bunch of little pokes from the stems of the alfalfa stubble. I never went to the doctor's office or hospital. Dad checked me over before he started milking and declared I would be okay. I didn't get to ride the rack for the rest of the year.

I loved to help or be with my dad. One spring day, Mom drove me to Carol Watts' house to play. They lived on a dead-end dirt road with very little traffic. At three o'clock Mrs. Watts sent me home. On the way home I spied my dad sowing oats in a field near the road. I climbed the fence and walked over to see if I could ride with him on the tractor.

To my dismay, Dad said he had too many things to worry about while driving the grain drill. It wouldn't be safe for me to ride today. Dad said I could stay in the seed wagon and wait for him until he finished and then ride home.

By four, Mom became worried. Where was her little Bobby? She called the Watts. Mrs. Watts sent Carol's older sister out on her bike to see if I was in the ditch or hurt. They didn't realize I was with my dad.

Now Dad had no knowledge of where I had been or where I should be. When he was finished and it was time to come home, he had an idea. He was driving the H Farmall with the drill. The B Farmall was pulling the seed wagon. He would have to come back for it later. He asked me if I wanted to drive the B home.

Now, I was only six years old. Dad sat on the little tractor, put it in low gear, and started me home. He drove along side with his tractor as we crawled to the house.

When we reached the house, he signaled for me to push the clutch forward. The tractor stopped. Dad ran over and threw it out of gear.

By this time, Mom had spotted us creeping across the field. She was relieved to see me and upset at Dad for letting me drive the tractor. She quickly called the Watts and told them Bobby was found. This was the beginning of my life with tractors.

My father seldom scolded me. I only remember twice. Once I was hanging around while he was repairing something out by our garage shop. I was swinging on the side extensions of the truck box bed and repeating, "Darn you, little devil." I suppose I had heard one of the men say this. I thought there was nothing wrong with repeating the expression.

Dad looked at me sternly and said, "Bobby, we don't talk that way."

That was all it took. I never used the expression again.

The second time I was spending my afternoon fishing for minnows with the Van Nice boys in the creek below their house. We did this quite often. We'd catch the minnows with grasshopper legs, then throw them back. Once in while we'd snag a bullhead, but they were never big enough to keep. This particular day, time slipped away. We were having fun. Jack was the only one with a watch. I had him check the time. He said it was only 4:30.

"Gosh, I must be getting home. I need to get the cows," I exclaimed.

I reeled in my fishing line and started running for the road. We always parked our bikes in the ditch. I was way ahead when Jack yelled out, "Wait up! What's your hurry?"

I didn't stop running and called over my shoulder, "I told you I have to get the cows in."

Jack and Joe had no chores. They really weren't farm boys. They were city boys who moved to the country. Their dad worked in town. They had no idea how important my job was.

I made it to my bike, laid the fishing pole on one of their bikes, and took off for home pedaling as fast as I could go. When I reached our place, I could see the cows were already penned in the lot.

Dad was by the machine shed. I slid to a stop and was going to explain, but before I could say anything, he scolded.

"Bobby, you know you are to bring the cows home. When I got home the cows were way out in the pasture. I had to get them. Next time you come home earlier from your playing with the Van Nice boys."

I hung my head and said, "I'm sorry."

"Okay, but don't let it happen again."

He returned to the barn. I ran to the house in tears. I explained to Mom I came right home as soon as I discovered what time it was. I didn't mean to be late.

Mom gave me a hug and said, "I'll explain what happened to Dad. Next time you take his old pocket watch. That way, you will know what time it is."

From then on, I took Dad's pocket watch with me. I was never late again. I carried that watch for many years. The watch is an heirloom, given to Dad when his mother died. Inside the cover it is inscribed with her birthday and the day she died. The watch itself has some value, but its value to me is priceless. Cheap wristwatches were not in vogue yet.

I remember one other episode where I was not on my father's good side. When I turned ten, I received a Daisy BB gun. It was a pump gun, so you had to pump it a couple of times before each shot. The only birds it would kill were sparrows. I did bang a pigeon once. I think I was lucky and hit it in the head.

Anyway, I spent hours walking around hunting sparrows. One day I discovered the haymow was full of sparrows, starlings, and pigeons. All were on the shooting list, because robins, goldfinches, and other good birds were off limits.

I wandered around the haymow taking pot shots at the birds. Some would fly up to the high windows and try to escape. They made a good target. I shot and shot. One pigeon hit the window with a crash, and the glass pane broke. The bird flew out.

I quit shooting. What was I going to say to my dad? I knew he would be angry.

I didn't say anything for days. One morning I came to breakfast and Mom said, "Your dad is very upset with you. He discovered several panes of glass broken in the haymow windows. Others were full of BB holes. He thinks maybe you shouldn't be shooting your BB gun."

At first, I was shocked, but I knew I was in deep trouble. I didn't deny my actions, but Dad's disapproval scared me. I started to cry. I told Mom I was sorry. She suggested when Dad came in for breakfast, I offer to pay for fixing the windows.

I waited for Dad to arrive. I was shaking. When Dad came in, Mom spoke first.

"Bobby said he would pay for the broken windows from his allowance."

Dad was quiet for a moment and said, "It isn't the cost of the windows that is bad. It is Bobby not being careful with his BB gun. Those windows are high up and I won't be able to repair them until next summer when the barn is full of hay. I think our boy should not be allowed to shoot his BB gun for three weeks and never in the haymow again. Do you understand?"

I meekly replied, "Yes, sir, I understand."

Believe you me, I was so relieved to still have my BB gun. My dad fixed the windows the next summer on the side of the barn which everyone could see. The windows on the west side weren't repaired for several years. Maybe he left them as a reminder.

Battle Scars

FARM BOYS TEND to receive a few cuts and bruises. It was my independent nature to do things myself. My dad and mom were always busy, and I didn't dare bother my sister with my trivial problems. At least, I thought I should not. She did come to my rescue many times.

My first scar I gained when I was very young, and don't remember exactly how it happened. I was with my father helping feed the hogs.

Dad fed hogs by scooping ear corn from the corncrib and throwing it out on the concrete floor. It was early November, and the crib was full. Dad was scooping corn directly from a wagon. He helped me into the wagon and had me sit high on the corn. I could throw individual ears to the hogs.

Somehow, I must have slid down the pile. Dad swung the shovel full of corn. He didn't see me. The edge of the scoop shovel hit above my eye on the right side of my head. That's all I remember.

I suppose they rushed me to Dr. Schroeder's office in Walcott for stitches. I do know I have a dimple and scar in my head to this day. One of the reminders of my childhood.

Dad fed beef animals for several years. Currently dairying was just one phase of his operation, and it was just twelve cows. Dad hadn't remodeled the barn yet. It still had stalls for the work horses. There was a center aisle for cows on one side and horses on the other. Hay was forked down a chute to the aisle. When the dairy barn was remodeled, it was built just for cows, since we no longer had horses.

Outside Dad fed beef steers in long feed bunks made of wood. For convenience, Dad placed the bunks close to the barn where the feed room was located. He would carry the feed, which was ground corn cob meal, in bushel baskets. The steers would come when he called. They licked the bunks clean. Their rough tongues sanded the bottom of the bunks until they were smooth as glass.

The bunks were placed on a slope. When the steers weren't eating, the bunks were good for sliding on your shoes. Mary and I liked to slide down and walk back to the other end to repeat again. One time, I either fell or decided to slide on my knees. A sliver of wood caught my overalls and tore them at the knee.

Mary said, "Mom's going to be mad."

I knew she was right, but my knee hurt. I could barely walk to the house. Mom scolded me and had me change my clothes. When I removed my pants, I saw a big, ugly black splinter. Now, maybe on an adult it wasn't so big, but I was only four or five years old. Mom had me sit on her lap as she tried to extract the splinter. She pulled with tweezers, but the piece of wood would not budge.

In today's world Mom would have taken me to the ER or urgent care. But instead, she said, "We'll wait until Dad comes in. I'm sure he will fix it."

It was summer and Dad always was late for supper. By the time he arrived, my knee really hurt. Mom had me lie on the couch and put ice on it. Dad looked at the monster splinter and said he'd work on it after supper.

After supper, he tried to fetch the splinter with a big needle. That didn't work. I tried to be brave, but it hurt.

Dad set me on his lap and pushed with his finger. The splinter moved a bit, but as soon as he grabbed it with the tweezers, it would break or slip. He worked for about an hour. My sister felt sorry for me and couldn't stand my crying. She went upstairs. Dad's last resort was a little pocketknife. He popped out the smallest blade and poured a small amount of alcohol on it.

He looked at me and said, "This might hurt some. I'll try to be quick."

Mom held me on her lap and Dad stretched my leg out over his knees. He slit the skin where the splinter was stuck. The wooden monster popped forward. He grabbed it with the tweezers and jerked it out. Now I was crying because it felt so good to have it out. Mom wrapped the cut with gauze and tape. I was given an aspirin and dad carried me to bed. My leg healed, but the scar is still there. Just one more badge of honor for surviving a childhood accident.

Another time, Mary was going to sharpen or pretend to sharpen something on the old pedal grindstone of Uncle Hen's. It had a seat, and Mary's legs were long enough to push the pedal and make the grindstone turn. Her problem was a fifty-five-gallon drum of engine oil which sat too close to the grindstone. She needed it moved a few inches.

I happened to be close by.

"Bobby, help me move this barrel. I think if you help, we can move it by tipping one side and spinning it to the right."

We both got on one side and began to push. The heavy drum moved a little. Mary always wore shoes. I was always barefooted. As we tilted the drum and began to spin it, my bare big toe went underneath the rim. The weight of the barrel overcame our strength and the fifty-five-gallon drum sat right down on my big toe. I screamed bloody murder.

Mary had an adrenalin rush and tipped the barrel enough for me to move my poor toe. I ran or hobbled to Mom. By the time I reached the house, my toe was already turning blue. The blood was building up under my toenail. It hurt. Mom quickly put my foot in ice water to help the pain. She had me try to walk, but it was too painful.

She said those famous words: "You'll have to wait until your father comes in. He'll know what to do."

Dad was baling straw at the neighbors. Thank goodness, he came home early. He was fueling the tractor when Mom appeared with me in tow.

"Bobby smashed his big toe. He's in a lot of pain. Do you think I should take him to the doctor's office?"

Dad finished the fueling and said, "Let me look at it. I'm going to Dipple's to finish baling his straw. He hasn't got very much."

Dad had me climb onto his lap on the tractor. He stared at my blue toe. Without a word, he reached into his pocket and pulled out his big pocketknife. Doctor Dad was going to operate.

He squeezed my big toe and began to drill into my toenail with his leather punch blade. I suppose it hurt, though I don't remember. Soon he drilled through. Blood squirted out like a fountain.

My toe immediately quit hurting. Dad wiped the blood away with his big red handkerchief and lowered me to the ground.

"Better soak that toe in hydrogen peroxide and wrap it good," he told Mom. "I'll be home by seven. Wayne is going to do the milking."

"Remember we are to go to Ellery's birthday party tonight," Mom reminded him.

"Yes, I remember. We will get there."

That evening we went to the Watts's for a birthday party. I couldn't wear a shoe, so Mom got one of Mary's old socks. She wrapped my toe and slipped the sock on. That evening I couldn't play with the other kids, but my toe survived. I lost my toenail for a while, but there were no scars, just memories.

If I needed to repair a toy or build a boat or oil a bike, I did it myself. I knew where the tools were.

One time I needed a screwdriver from Dad's shop. They hung on a rack above the workbench. I was a little short to reach the one I needed. I climbed onto a bucket, but it still wasn't good enough. I had discovered earlier I could climb up the end of the bench via some cross members. Dad never had an actual toolbox, his tools just lay on the bench. I crawled across the tool-filled bench, retrieved my screwdriver, and was backing down when I felt a sharp pain.

I looked down at my bare foot and there was a hand sickle sticking in my foot, just below my ankle. I didn't hesitate and pulled the sickle out. My foot was bleeding all over the bench. I know I didn't cry and holler for Mom because she couldn't hear me. I wasn't supposed to be on the bench in the first place.

I climbed to the floor, found an old rag to wrap my foot, and limped to the house. Mom didn't know if she should scold me or be happy it wasn't any worse. She cleaned the wound with a washrag and again used the ever-present hydrogen peroxide. I probably should have had some stitches, but the doctors' office was in Walcott and Davenport was the nearest hospital. I healed in a few days with my foot wrapped and shoes on. I was left with two ugly scars on my foot, just one more honor badge.

As a kid on the farm cuts and bruises were common. Yes, I received many more in later life, but these were the ones with the most vivid memories. Luckily, I never lost a digit. Many farmers lose a finger or hand by being careless or in a hurry. In some cases, a man loses his life. Farming is considered one of the top ten most dangerous occupations. It also can be one of the most rewarding.

My Pets

WALTER KEMPER CAME every morning in a refrigerated truck to load our filled milk cans. He also left ten empty cans. He liked our stop because it was midway through his route. He could take a break and rearrange the cans in his truck for the rest of the day.

One day Dad called me to the barn after Walter had left. He said, "Walt left you a present."

Dad handed me a little kitten. She was striped with four white paws. I immediately fell in love with this kitten. I named her Boots. She stayed and had many litters of kittens.

She always had her kittens in the haymow. When I noticed she didn't come for milk for several days, I knew she had had her litter. I waited until she returned to the mow, then I followed her. I became an expert on making little mews. The new kittens, thinking it was their mother, would emerge from a hole between some bales of hay. They had made it through the tom cat period.

Toms kill baby kittens before their eyes open if they find them. I never understood why. Dad said it was the way nature kept the numbers down.

Boots lived to be thirteen years old. She was a great mouser, but in later years she lost some of her teeth. I found her lying by the corncrib one day. Age had caught up with her. I buried her behind the barn.

I had one other short-lived pet, another kitten. I received this one from my Grandmother Shepard. I named her Pug. She was a gray semi-long-haired cat. Our barn cats had the bad habit of sleeping on our cows' backs or curling up in the warm straw where a cow had laid. They had to be quick if the cow decided to lie down again. This is what happened to Pug. She didn't move quick enough, and a twelve-hundred-pound Holstein cow flopped down and flattened Pug like a pancake. I'm sure it was a quick death. This is something farm children get used to, life and death.

I had one dog. We got her during World War II. Dad went to an auction in Muscatine where all proceeds went toward buying war bonds. Businesses and people donated items to sell. Dad came home with a Quaker oatmeal carton with holes punched in the top. He gave it to me and said, "Open it."

I opened the round carton and inside was a brown and white puppy. I named her Lassie because she had the markings resembling Lassie, the movie wonder dog.

My Lassie was small and short haired. Her only talent was to howl when someone honked the car horn. Her other claim to fame was when our family returned home from a vacation, Lassie would run around the house at full speed until I stopped her.

Lassie lived for twenty-some years. One day, my mom called while I was away at college. She said she found Lassie at the bottom of the basement stairs. She evidently had a heart attack before trying to climb the stairs. Our hired man buried her.

Hazel Dell #3

MOST COUNTRY SCHOOLS started the day after Labor Day and ended the year in early May. We took very few days off for holidays. The first big holiday was Thanksgiving. Next came the Christmas break, which normally was from December 23 to January 2. There were no other breaks except Good Friday before Easter.

We acknowledged Columbus Day, Veterans Day, Valentine's Day, Lincoln's and Washington's birthday, but never had a day off; thus, the required 180 days of education were accomplished before May fifteenth.

When I was old enough to attend school, I attended Hazel Dell #3. Our school was half a mile away. It was one of three rural schools in Montpelier Township where we lived. It had just one room with three sets of desks, each set for a different size of student. The desks were attached to a flat board, four to five in a group. The teacher could move them around to suit her needs.

The door was in the front of the building. It was protected by a small room or vestibule. This was where the muddy boots were stored. On the wall just inside the door were four rows of hooks for hanging coats and hats.

Hazel Dell was more modern than Patterson and Pine Mill, the other two schools in the township. We had a coal-fired furnace located in the basement of the building. The other schools had their heating unit in the middle of the classroom. If you can picture this 20 x 32 room with rows of desks, a teacher's desk, a large cabinet holding two sets of encyclopedias, two rows of hanging lamps, and a huge four-foot by four-foot heat register directly above the furnace.

The stairs leading to the basement were covered with a large trap door. It was heavy and only the teacher or the older students could raise it. The basement wasn't more than a dug space beneath the school building. The floor was dirt. There was a small stack of kindling for starting the fire in the furnace and a larger bin containing large lumps of coal. The teacher was expected to start the fire, keep it lit, and bank the furnace at night before she'd go home. Banking the furnace meant arranging the lumps of coal around the edge of the firebox so they would burn slowly and not be out by the next morning.

This worked overnight, but on weekends was seldom successful. One of the fathers would come early and start the fire in the furnace before school. There were weekends during the cold months the fire would die. Everything in the school building would freeze. Our white paste and ink would be frozen solid. We would sit around the big register in the front of the room in our coats and mittens while teacher read to us. In an hour or so, the room would warm enough, and we could return to our desks and hold classes.

I was in seventh grade when the school board installed a coal stoker on the furnace. This machine fed small pieces of coal into the firepit automatically. The coal stoker bin was filled maybe twice a week, according to the weather, but it eliminated the chore of banking the furnace each evening. Unless the stoker bin became empty, there was always heat in the building and it was controlled by a thermostat. Filling the stoker became the chore for the older boys, of which I was one.

Our chore would be to shovel coal into the automatic stoker. The other part of the job was cleaning out the clinkers and ash. There is a story about Jack Van Nice and I roasting a dead mouse while we were cleaning out the clinkers, but I'll save that for later.

The front of the school room was covered with black boards made of slate. They ran across the front and around the side walls for several sections. I believe there were eight different sections, which was many more than most one-room schools had. Each had a tray for chalk and a felt eraser. On Fridays of every week, one of the chores performed by the students was washing the blackboards and pounding the erasers together to rid them of the chalk dust.

The chalk all came in boxes packed in sawdust. Some of the teachers would allot one piece of chalk to a student; others just limited the chalk available. We were all taught to use our chalk sparingly. Most items at Hazel Dell were sparse. The only time we received paper from the teacher was when it was for an art project.

As for plumbing inside the building, there wasn't any. The girl's and boy's restrooms were outside privies. For some reason, the girl's privy was located a good fifty yards south and west of the building. The boy's facilities were much closer, located thirty feet behind the school. They were bitter cold in the winter. I don't know how the girls stood the inhumane cold wooden seats, but that's how it was. We didn't know any different. Some of the families still used outdoor restrooms at home.

Our water supply was a hand-dug, brick-lined well just north of the building. It had a pump with a long handle to get water. We would fill a big stone crock and two big kids would carry it inside the school room during the winter months. When the temperature permitted, the crock was stored in the vestibule.

The crock had a small push button spigot to fill our cups or glasses. Each student was required to bring their own drinking cup. We'd draw our cup of water, drink it, and place it upside down on a shelf. The shelf was just a painted board hanging on the wall. Once one of the moms cut a piece of oilcloth to cover the shelf. This made it possible to wipe the shelf clean once a week. Maybe it wasn't the cleanest, but I don't recall anyone becoming ill from the drinking water.

The first day of school, the well and pump had to be primed. The leathers dried out over the summer, so water was slow in coming at first. If we'd pump fast enough, the water overtook the loss and start flowing out the spigot. Occasionally a field mouse would make the dry chamber her home. At first, the mouse would stay put inside the water chamber until a gush of water finally made it to the top. The poor soaked mouse would come out of the spigot and run for its life. With so many children around laughing at the scared animal and trying to stomp on it, the mouse seldom had a chance.

Once the mouse had been evicted and the water looked pure, we just pumped a little longer to clear out any nesting material. From then on, it was good drinking water. Before I became a student at Hazel Dell, everyone drank from the well which was covered with wooden planks. The excess water would drain through the cracks and down into the well. We recycled our water and didn't realize we were conservationists. We also washed our muddy boots at the same well. The muddied water flowed through the cracks between the boards. Everyone was fine with the setup because most wells at home were built the same way.

One summer our dads rebuilt the well platform. Maybe the old boards were unsafe, or maybe an edict came from the superintendent's office, but the well was covered with a concrete platform. This did stop contamination of the well below, but the occasional mouse continued to appear.

Our water was considered safe until the Van Nices arrived from the big city of Davenport. Mr. Van Nice thought we should have the water tested for impurities. Of course, the test showed the water was impure for drinking, although we'd been using the well for years.

What purifies water? Chlorine. Mr. Van Nice dumped several chlorine pellets into the well. I don't how many pellets he threw into the well, but it was undrinkable. The first water pumped was a yellow terrible tasting fluid. We pumped and pumped until the water was clear. It didn't make any difference; the water tasted icky. To solve the problem, first thing every morning we had to pump the well. The rule was developed: pump until clear, then add one hundred more pumps. By October the water was drinkable. As far as I know it was never tested again.

Hazel Dell had other discrepancies. Although we had electricity, we didn't have a telephone. If a student became ill, they had to stay at school until the end of the day. Their parents could not be contacted.

If there was an emergency, the teacher or older student would walk or run to the Dipple's home. Henry Dipple lived less than a block from the school. Mrs. Dipple was always home. I remember only one or two times we trekked to the Dipple's for the emergency call.

Our schoolyard was the mandated one acre in size. The problem was the yard was split by a hill. Two-thirds of the yard was near the school building and somewhat level. The remaining part was twenty-five feet lower in a slough.

Neither lent itself to a baseball diamond. The uphill diamond's outfield sloped down and away. We tried several configurations, but none were ideal. It was always trees, road, or no backstop. We could use the schoolhouse for a backstop, but there were three huge windows on each side. An errant ball might smash through the glass.

The lower diamond was limited by a board fence on one side, a steep hill in the outfield, and an area of gopher mounds near third base. The baseline between first and second skirted by a washout filed with junk. It was definitely a hazard. Being rural kids, we adjusted our game rules according to the layout. A ball hit over the right field fence was a double. A hit into the washout was a single. Third base was an uphill run and a downhill run to home. Home plate was either a board or a burlap sack. The bases were sticks or wooden scraps.

One other piece of playground equipment was a merry-go-round set on a slight angle. The angle allowed the law of physics to work. Once started, if you hung on the downhill side, your weight shift would cause the wheel to spin faster. The more you pumped or moved in and out, the faster it would go.

Some of the bigger boys were quick enough to shift their weight to spin that old merry-go-round at a good speed. The spinning force was such once you were pinned against the grasp bars, you could not move. No one ever got hurt except Harold Paustian. Once and only once he decided to sit on the top of the center post while the merry-go-round spun. It was all fine until he discovered the plate on the top moved with the spinning equipment, but the grease fitting remained stationary. The fitting ground a neat hole in the seat of his pants. I imagine he had a sore spot on his butt, too.

The last piece of playground equipment was a basketball hoop. The hoop was mounted on a four-by-five bank board attached to a tall pole. The hoop had no net or any way of attaching a net. The court was grass and rough. When one would try and dribble, the ball made up its own mind and went everywhere. It was a passing game. When someone did take a shot, the backboard would shake back and forth.

Other schools had several swings mounted on steel supports. They were sturdy and strong. Ours was a swing hanging from a large limb in a walnut tree. It lasted until the rope deteriorated and broke. It wasn't replaced since no one was brave enough to climb the tree and re-tie a new one.

My Teachers

THE FIRST-YEAR CLASS at Hazel Dell was deemed primary. It would be the same as kindergarten today. I was the only student in the class and Miss Mary Kemper was my first teacher.

Miss Kemper had her quirks. She let the students have a long rope. We'd play our games and if we were in the middle of an inning, she'd let us finish. If the snow was good, she would allow us to take short morning and afternoon recesses to lengthen the noon recess. She was a farm girl with eight siblings and understood our time together.

A couple times a year we'd get too rambunctious, and her rope would shorten. We would be punished for whispering during class time. She tended to pull this stunt the night of some other important event, such as our annual Halloween Costume contest at the Farm Bureau meeting. She made a list of names on her desk. Every time she caught someone whispering, she'd mark an X. Each X was for five minutes after school. Usually, about two o'clock, someone realized the ploy. The word quickly spread, but it was too late. Our goose was cooked.

One day I was a good boy. I only received two marks. Carol Watts did the best with one X. It was my sister Mary who was caught the most; she had 6 Xs. This meant she had to stay after school for thirty minutes.

Of course, I had to stay there also because I was not to ride home without her. Mary was furious because her best friend Joyce Heuer was only marked for ten minutes. Mary claimed Joyce whispered as much as she did, but she didn't get caught. We hurried home, did our chores, practiced piano, and got ready for the costume party. Miss Kemper would do this again sometime in March or April. By then we had forgotten about the Halloween caper, and most of us would be caught again.

One spring day Miss Kemper became ill. It was mid-afternoon. Miss Kemper quickly got up from her desk and headed for the door. About halfway there she vomited and passed out. Luckily, she sat down in one of the desks before falling.

Now, we had no telephone and no other adult. Did we panic? Never! Mary and Marilyn Watts headed for Dipple's to make a phone call. One of the eighth graders, Bobby Noll, got the first aid kit.

Inside the kit he found some smelling salts. He gave Miss Kemper a whiff. She immediately revived and was thoroughly embarrassed. One of the older girls helped Miss Kemper to a bench on the north wall where she could lie down. Another couple of older girls cleaned up the mess. There was no janitor.

We stayed around for about thirty minutes until Miss Kemper's mother and father arrived. I believe my mom also arrived because she dismissed school. Any students who were walking, Mom drove home. The next morning Miss Kemper was back on the job, bright and cheery. She told us later she must have had food poisoning.

I had Mary Kemper for my first two years. The second year Miss Kemper had a boyfriend, Wally Hauck. He worked for the railroad. Some days he was off and would drop by school around noon. We loved him because he would play ball. The best thing was, he would go into the school building during recess. He kept Miss Kemper occupied and our noon recesses sometimes were thirty minutes longer than usual. We never caught them kissing or anything like that. We just kept on playing and didn't bother to go inside.

On his days off, Wally worked for my dad. He loved to drive tractor and do field work. This meant Mary and I had a special relationship with him. He would stop at school after dismissal to pick up Miss Kemper and sometimes give Mary and me a ride home. Later that summer, Miss Kemper got married and became Mrs. Hauck.

My second grade started with Miss Geraldine Hidlebaugh. She was the picture of what a teacher should not be. It didn't take long before we knew how to work the system. Since I was in second grade and Allen Paustian was in first, we were playmates. Carol Watts was in fourth and playing children's games was not suited for her mentality.

Miss Hidlebaugh did not want to deal with little boys. Many days, Allen and I would be dismissed at ten in the morning and play outside the rest of the day. The only rule was to not go outside the schoolyard and stay below the hill so we wouldn't disturb the rest of the pupils.

Of course, Allen and I obeyed everything but staying inside the school boundaries. Dipple's pasture provided many more opportunities for entertainment. We played in the little creek just beyond the fence. We were close enough that we could quickly climb back into the schoolyard when Miss Hidlebaugh called.

Inside the school, study was slowly falling apart. My

sister was frustrated because Miss Hidlebaugh didn't hold class as scheduled. One day Mom arrived to pick up Mary and me for piano lessons. I don't believe I was taking lessons yet, but I went along anyway because no one would be home for me after school. This day Mom pulled into the school drive only to see her son coming out the window.

You see, Miss Hidlebaugh ignored her second grader. I had the last desk in that row and if I was careful, I could open the big door on the library, which was just an enclosed bookcase, and sneak out the open window. I was sneaking back into the classroom thinking I was pretty smart until Mom showed up. Boy, did I get a scolding about that, and not from Miss Hidlebaugh. I don't think she ever suspected anything was going on.

It wasn't long before the school board contacted the county supervisor, Mr. Ralston, and Miss Hidlebaugh was dismissed. To take her place, Mrs. Hauck returned to teach.

The only problem was Mrs. Hauck was pregnant. It didn't bother any of the students, since we'd seen pregnant women. I don't think it bothered our parents, but it was against county rural school policy. A pregnant woman might miss school days or be unable to lift the coal into the furnace. She was considered a liability or handicap.

I don't know how the school board convinced the superintendent to let them hire her, but there probably was no other option. Mary filled in from Thanksgiving to the end of February. At that time her sister, Martha, would graduate from college. At that time Iowa State Teachers College was on a quarter system and Martha would graduate in February.

Mrs. Hauck realized our school progress was far behind compared to the other country schools, so she did her best to help us catch up. We were so far behind that during the week of Christmas break, we had to return to school between Christmas and New Year's. Our Christmas holiday was only three days long. It didn't really bother most of us. We had more fun being together.

By February we were caught up and Martha Kemper, Mrs. Hauck's sister, took over. She taught the remaining days of that year and the next two years.

Miss Martha Kemper was great on field trips. We didn't go to zoos or museums. She would receive permission from the neighbors to take the students on a hike through their pasture and woodlands. The hike would start right after lunch. She knew a lot about plants and animals.

One warm day in the spring, we had the windows open. Suddenly, we heard a lady's voice crying, "Help! Help!"

At first, Miss Kemper thought it was one of the students playing a trick on her. But there it was again. "Help! Help!"

She went to the window and saw Mrs. Dipple coming up the road waving her arms as she ran.

Miss Kemper hurried to the door. After speaking with Mrs. Dipple, she returned and said, "I'm going with Mrs. Dipple. Her husband is sick. Marjorie, I'm leaving you in charge. The rest of you just continue with your lessons."

Marjorie was the oldest and smartest of any of us. She went to the front of the room and took Miss Kemper's chair. She gave us a stern look and said, "Everyone continue. Third grade, come up for reading."

Miss Kemper returned about thirty minutes later. She sat down at her desk and sighed. Her facial expression was sad.

She said, "Mr. Dipple died. When I got to their house, he was face down on the kitchen table. I knew he was dead. I called his son."

When I think back, she was very brave for a young woman of maybe twenty years old. To become an elementary teacher in rural schools, only two years of college were required. If you were young when you graduated from high school, it was possible to be only nineteen when you started teaching.

Over the summer between my fourth and fifth grades, Martha wed Herb Hetzler. In the fall, I had my fourth teacher.

The new teacher for my fifth grade was Doreen Henke. She was a tall, thin person with red hair and skin covered with freckles. I liked her very much. I accomplished things I thought were difficult. She made everyone read up a grade. Just before Christmas break, she asked the school board if she could have a few more days off. We students didn't know why, but when we resumed classes after New Year's, her name was Mrs. Fuller. She married an Air Force pilot who was to be deployed in January.

Except for her new name, Mrs. Fuller didn't change. I can't really pinpoint why she stands out in my mind except she was a teacher who challenged me to do better. Sadly, Mrs. Fuller quit over the summer and followed her husband. I was disappointed to lose her.

Years later I met her again at one of my book signings. She was still tall and thin. Her hair was red tinged with gray, and she still had those freckles.

Her comment was, "I'm glad some of my teaching wore off on someone." I was surprised she still remembered her student from Hazel Dell.

Before the next school year started, Hazel Dell had the luxury of a telephone installed. The phone sat on a small storage closet near the back of the room. It proved its value that year.

I was taught in sixth grade by the third Kemper sister, Doris. Her claim to notoriety is that she was the only teacher who stayed overnight at our home while a teacher at Hazel Dell.

It occurred one winter day. The weather changed during the day from a decent day to one of freezing rain and later snow. School didn't let out early like they would today, but our parents came to pick us up. We had a carload in our Studebaker, since Mom delivered kids who didn't have rides. Doris stayed until everyone was safely out the door. She had papers to grade.

I did my chores feeding the bucket calves and feeding the chickens. But it was becoming dark, and my sister had not come home from high school on the school bus.

Before Mary went to high school, most students from Hazel Dell either attended Davenport High or Muscatine High. In order to attend, students would stay in town with a friend or relative until they were old enough to drive. They could receive a student's permit at age fourteen and could then drive to school. This meant purchasing another car and maintaining it.

My dad and several other dads from other rural schools in Montpelier Township decided to organize a school bus. The farmer who agreed to run the bus was

Mr. Edgar Kemper, the father of all my Kemper teachers.

On this particular day, the ice storm hit. City school officials didn't realize the treacherous road conditions out in the rural areas. They didn't dismiss school early or release those riding the Kemper buses. I was practicing my piano lesson when I heard Mom talking on the phone.

"Okay, I'll call the parents on the Davenport phone lines," she said.

"What's up?" I asked.

"That was Beulah Kemper. The school bus Mary rides can't make the hills on the gravel roads because of the ice. All the students living away from Highway 61 are staying the night at the Kemper's." At this time Highway 61 ran along the Mississippi River from Davenport to Muscatine.

I thought, "What fun Mary is going to have with all her friends."

The phone rang again. Mom answered, "Sure, why don't you come right away before the lane gets too slick."

Then she turned to me and said, "Doris is coming here for the night. Her father said it is too dangerous for her to drive home."

My mother referred to Miss Kemper as Doris because of our family ties with the Kemper family. Soon Miss Kemper arrived. This was an unusual situation. Miss Kemper was staying at our house while Mary was staying at Miss Kemper's house. I remember it was difficult to decide, do I call her Miss Kemper or Doris? Outside of school, Miss Kemper was a family friend. I called her Doris at church. I don't remember now what I decided.

The storm was bad enough that the schools were closed the following day. By late morning, Mary made it home and Miss Kemper returned to her home. I don't remember if we lost our electricity or not. I'm sure in a couple of days the ice melted, and school resumed. Rural schools seldom were closed due to weather because we all lived close and school buses were not an issue for elementary students.

I only remember one time when our school had a substitute teacher for a few days. It was so brief I don't remember her name. I only remember she was married and didn't play games with us. She basically just babysat.

The last two years at Hazel Dell I had Mrs. Laura Schroeder. The scarcity of elementary teachers for rural schools caused rules to be relaxed. Married teachers were allowed to teach. She lived in the village of Montpelier about five miles away.

Mrs. Schroeder was a good teacher. She also was the last teacher Hazel Dell employed. A year after I graduated from eighth grade, the rural schools in Muscatine County were consolidated. Our rural schools were closed, and students attended schools in nearby towns or cities. It was the end of an era.

The Student's Day

OUR SCHOOL DAY started at nine in the morning with the pledge to the American flag. The day ended at four in the afternoon. There were four periods to a day with a fifteen-minute recess at ten-thirty and two-thirty. The noon recess was one hour long. This gave us some serious time for games or sledding.

Each morning, most students would arrive at eight-thirty. We needed some playtime before school since we lived too far apart for playing after we got home. At nine, Miss Kemper rang the hand bell from the front door. We had five minutes to get in our seats at our desks. Since I was the youngest and smallest, my desk was in front of the others, right in front of Miss Kemper's desk.

Every morning she would read a chapter from a book such as *Tom Sawyer* or *Huckleberry Finn*. The story didn't attract my attention. I had a brand new Big Chief tablet and new pencils and new crayons. This was a new adventure for me.

As Miss Kemper read, I scribbled and drew on my new pad. I remember I was drawing a black cloud and it was raining. I drew a bolt of lightning and in a loud voice said, "Boom!"

All the kids looked at me, especially Miss Kemper. She kindly said, "Bobby, this is a quiet time for others while I read. You must be quiet, too."

I suppose I was embarrassed, along with my big sister. I know it was the last time I drew pictures while Miss Kemper read.

This was the start of my adventure through the nine grades of Hazel Dell #3. My class time with Miss Kemper was short, but almost always the first class each period. Each class would have their time with the teacher, starting with the lowest grade. We would sit on a bench either in front of her desk or to one side.

As the only student in primary, I worked on my ABCs and coloring skills. Most of my day was spent listening to the others do their lessons. Besides my new coloring book and Big Chief tablet, I remember I had a subject called phonics. It was supposed to be a precursor to reading.

I had a thin packet of worksheets which were purchased at a bookstore in Muscatine. They were glued together on the edge and were to last until Christmas break. I was finished before Thanksgiving. Miss Kemper acquired extra worksheets from the superintendent's office. They were adjusted for more advanced students. This doesn't say I was advanced, but I was older than most primary students. I was almost six years old when I started school.

Each teacher had her own style. We would present our work or read from our readers. Everyone started out with Dick and Jane, their little sister, Sally, Spot, the dog, and Puff, the cat. "See Dick run. Run Dick run." Every rural student remembers these primers.

I loved to read and was ahead of schedule in every subject. As I became older, I roared through the basic readers. By second grade I was reading at third or fourth grade level.

Our school had a very limited library. We had a few of the classics, but mostly it consisted of World Book Encyclopedia and the Books of Knowledge. The Books of

Knowledge were volumes of facts and writings. They were the oldest books in the library and of little value to me. I believe they were some my father used when he attended Hazel Dell.

Since the library was limited, the school had access to the public library in Muscatine. Our teacher would visit the library every other weekend. She would select books for all grades. If there was a special book we wanted, she'd tried to bring it back to school.

We had two weeks to read our books. If we were in the middle of a story, teacher could renew the issue for two more weeks. We were very careful with the Muscatine Library's books. For some reason, city people thought rural children were rough on books. This was one of the many fallacies, city people had of rural people.

There was one book I read several times. It was *Corn Gold Farm* by Paul Corey. It was the same story line in all his books, a story of a young person working hard and sticking to his ideals. In the end, his work ethic and ideals prevailed.

In this particular story, the young farmer was using basic conservation practices which at the time were very new to the farming community. Dad was one of the leaders of this practice. He was on the first conservation board of Muscatine County. If my dad thought it was right, it had to be right. How could my dad be wrong? This book spoke to me about the values of conservation.

My best memory of the reading period was listening to the upper grades read. Their readers had stories about Paul Bunyan and Babe, the Blue Ox. Gulliver's travels always fascinated my curiosity. By the time I reached sixth, seventh, and eighth grade I knew all the stories. I had heard them several times before; in fact, I had read them much earlier since Mary had her reader available. Reading these stories for the class was merely academic.

The first four days of the school week were basically the same. Friday was the change in subjects. Instead of arithmetic, we studied health. Second and third periods varied, but fourth was art or music. My first few years we used an RCA Victrola which we had to crank to play. After getting the record spinning, you would gently place the pickup arm on the recording. The sound was scratchy and full of static. The needle which ran in the grooves on the disc sometimes would become dull or worn. The arm started to skip across the record. A teacher or an eighth grader would take a small screwdriver and loosen the needle and replace it with a new, sharper one.

One year, someone at the superintendent's office decided old wind-up Victrolas were obsolete. The school purchased an electric record player. It was encased in a brown wooden cabinet and had a lid which opened from the top. We thought we were on the cutting edge of technology. To top it off, we had records which came in an album containing several recordings. Each record had songs for a specific grade level.

During the year, we were required to learn most of these songs by memory. The records had bands on the disc which contained one piece of music. Of course, we had simple songs like "Farmer in the Dell" and "Mary Had a Little Lamb." The older students were to memorize "America, the Beautiful," "Waltzing Matilda," "Jeanie with the Light Brown Hair," and many others. If you were good enough at memorization and a seventh or eighth grader, you were eligible to participate in the All-

County Choral Group, which sang at the eighth-grade graduation ceremonies in May.

Art class was generally a craft course. Teachers attended seminars to learn new skills. If the art project required duplication, she used a device called a hectograph. This was way before Xerox and any such duplicating equipment. Town schools had mimeographs.

The duplicator was very simple. It was a tray full of a gel. Teacher would make a master copy with copious amounts of ink. She would place the paper on the gel and roll it with a small rolling pin. After doing this, she could place plain white paper on the gel, roll it, and have a copy.

This would provide maybe ten to twenty copies. The number of copies was determined by the quality of gel in the pan. Eventually, the gel would become so impregnated with ink it no longer would copy. Teacher would take it home, warm the pan to melt the gel, pour out the old, and replenish it with new gel. The life of the gel depended on the teacher. Some loved art, for others, it was just a necessary class.

Miss Doris Kemper attended an art seminar held for teachers. She came back with a project I really liked. She showed us how to draw spring flowers. She would show us each stroke and color. Soon, we had drawn a lilac, tulip, or crocus. People like Carol Watts excelled. She was always the best artist. Miss Kemper amazed us with how easy it was do art.

Friday art class was sometimes interrupted by birthday parties. If someone had a birthday during the week, their mother would arrive with cookies and drink. Sometimes we had time for a game or two. The last Friday before the school year was over, all the students who had birthdays over the summer would have a combined party.

Once in a while our teacher would attend a conference or seminar on a weekend or holiday. Miss Martha Kemper attended a science class on the subject of airplanes. The next Friday, she explained how airplanes fly and how they go up and down. She used a paper airplane to demonstrate. She had everyone fold a paper glider of our own. We made the paper planes fly in circles, loop the loop, and up and down. She encouraged us to color them and put our trademark on them.

Little did she realize she started a fad for a few weeks. All of us began to fold and color paper airplanes. During recess she allowed us to fly the gliders in the school. We had contests to determine which one would travel the farthest. Miss Kemper realized fads quickly fade into something else and as soon as the weather turned nice, we'd be outside.

We always had a Valentine's Day party, too. Everyone stuffed their Valentines into a big box that was decorated with hearts and lace. Usually, it was the task of someone in the upper grades to construct the box. When the time arrived for the party, some mothers would arrive with cookies and maybe a drink. The teacher chose a student to open the box and hand out the small cards. We read our cards and giggled. We all knew we loved each other but not in the way the cards presented it. After the cookies and drink, teacher would hand out her Valentine gift, which was little candy hearts. We'd giggle some more while we read the printing on each heart.

The grass in the schoolyard was cut once at the beginning of the year in September. A neighbor would mow it with a sickle mower and bale the grass to feed to his cows. One year, the school board decided to let Mr. Dipple pasture the yard. The only problem was, the cows left deposits of cow chips. The week before school started, a group of dads walked the yard picking up cow chips.

In the spring, before the green grass started to grow again, the teacher would have a lawn maintenance day. Over the winter sticks were scattered around and there was a lot of old dried grass. Every student brought a yard rake, and we all raked the school yard. By the end of the day, we'd built a large pile of grass. Just before school let out, we'd light the pile and have a bonfire. The teacher would provide marshmallows to roast as a reward for our hard work.

Grass burns hot and fast. It isn't the best for roasting marshmallows, but we did it anyway. Some got a little ash on their marshmallow, but ash won't make you sick, so you'd just brush it off and pop the marshmallow into your mouth.

The most exciting day of the year was the last day of school. It was celebrated by the whole community. We students arrived at our normal time to start our day, but that was it. The morning was filled with washing the blackboards for the last time, sweeping the floors, and stacking books away. Our desks were emptied, and all unnecessary items were thrown out or packed for going home.

The teacher handed out our report cards. At the bottom of the card was a statement that we had graduated to the next grade level. We all knew we were moving to the next grade, but it was still exciting to see the message in writing.

I only remember one girl not moving on. She was required to take a grade over. I remember feeling sorry for her. She came from a home where she had little help from her mother or grandmother. Her parents had divorced. Today there would be special education teachers to help. Fortunately, when I met this girl many years later, she had a family and was doing fine.

Just before eleven, our mothers arrived with a picnic lunch. The teacher's desk plus a couple of card tables stretched across the front of the room and were covered with a potluck of food. At noon the dads started to arrive. It wasn't just the fathers of the current students, but the men of the community. It was a time to get together and visit. Most were farmers and took time from their field work or other duties to attend the celebration.

The most fun was after dinner when we played a pickup baseball game. The dads joined in. They'd smack that softball a mile. I always felt so proud when my dad was up to the plate. He generally got a hit. The dads all took a turn at batting before they returned to work. The students continued the game for a while. The mothers chatted with each other and put the dishes away. Those students who rode bicycles had to ride home. Others got a ride home with their mothers. I suspect the teacher heaved a sigh of relief as she locked the door for the last time.

In 1949, my solitary school life changed. Dad and Mom decided to build a new house on our farm. Instead of tearing down the old one, Dad sold it to a man from Davenport. That man had just purchased a farm about half a mile away. His farm was very run down. The house was in poor condition and small. I don't have a clue how the two men found each other, but the man bought our house and moved it down the road. His name was James Van Nice.

He and his wife, Katherine, had two sons, Jack and Joe. Jack happened to be my age and became my classmate for the rest of my stay at Hazel Dell #3. We also attended the same high school and college. His brother, Joe was two years younger. From third grade on we had a friendly rivalry. I liked Jack as a playmate but wasn't so fond of him horning in on my private class.

Jack and Joe formerly lived in the largest city in eastern Iowa, Davenport, before they moved to our neighborhood. They had experienced playgrounds and close neighbors. Their friends in town had places to go and play baseball, ride their bikes, and generally hang out. They had a great let down coming to rural Hazel Dell.

At first, both boys indicated that they were better in every sport. They probably were, but we at Hazel Dell didn't always play baseball. We had other games such as May I?, Andy, Andy Over, Johnny Cross Over the Ocean, and Stealing Sticks for playing outside.

At first, I tolerated Jack. He was much better than I was, but I wasn't going to let this city slicker outdo me. This friendly rivalry became more and more competitive. I don't remember why we argued, and I don't remember who threw the first punch, but one time during afternoon recess, Jack and I had a knock-down, drag-out fight.

It wasn't punching so much as a wrestling match. We were going at it close to the batter's box on the uphill baseball diamond. Next thing we knew, Miss Kemper was pulling us apart. I was on top and proving my superiority. We both were scolded and sent inside. Both of us stayed after school. I don't know who got to go home first, me or Jack, but it was our last fight. We became and still are good friends.

When Jack and I were in sixth grade, there was no seventh grade, and Carol Watts was in eighth grade. She always had grades in the 90s. We didn't have letter grades in Hazel Dell. Jack and I strove to stay at her level. Reading and subjects like geography were easy subjects to keep pace. Arithmetic was our nemesis. Carol would rack up 95s and 100s. She was top in her class, though she was also the only one in her class.

Jack and I did our best, and the first eight-week report card showed we were keeping up. We could keep pace with that smart Carol. You see, our teacher, now the third Miss Kemper, allowed us to check our own papers when we had our class time at her desk. It was easy to "miss" a mistake. Our math grades improved. We were on a roll.

One fateful day, Miss Kemper made us grade each other's arithmetic papers. I found several mistakes in Jack's lesson, and he found several mistakes in mine. Miss Kemper never cracked a smile. She had caught us red-handed, but she never said a word. From then on, I checked Jack's papers and he mine. Our grades were no longer as good as Carol's. Lesson learned was you can't outsmart a good teacher.

When Carol Watts graduated and was no longer a part of Hazel Dell, Jack and I became the big kids. We held that honor for two full years. We also became the designated chore boys.

In the morning we'd carry the fresh water for drinking inside. At the end of the day, any water not used had to be emptied. Another task we had on our list was cleaning out the clinkers or ash from the furnace. The furnace was in the basement of the school. To enter the basement a heavy door in the floor had to be lifted and hooked. The stairs down were steep and open.

When the school used lump coal for heat, once a day someone would pull the ash grate open and the spent coal would fall through to the ash pit. From fifth grade on, once or twice a week, Jack and I would shovel the ash into buckets and carry it up the stairs, through the school room, and outside to an ash pile next to the only red cedar tree on the grounds. The clean-out depended on how much heat was needed to heat the building.

When the school board installed the automatic coal stoker system, the heat was controlled by a thermostat. It became our job to fill the stoker bin with coal. The coal was about one to two inches in size. On cold days, we did this every day. As the weather warmed, the filling lessened. The coal still left ash. Because of the system, there was no ash pit. Jack and I were to clean the hot ash and place it in pails.

Stoker coal, as it burns, melts into small clumps of hard rocks we called clinkers. Since the ash pit no longer existed, we used a long-handled poker and a tool which had a very long handle welded to a small frying pan. One of us would scoop the clinkers with the pan while the other held the ash pail. The best part of the job was it was away from the teacher's eyes. We'd mess around in the basement to kill some time. To look like we were doing something we'd straighten some of the junk which accumulated over the years. One day, while sorting through the junk, we found a dead mouse.

"Let's put the mouse in the ash scoop and watch him

fry," I said to Jack.

Jack agreed. He gingerly picked the mouse up by its tail and placed it in the metal ash pan. I stuck the mouse inside the firepit. We watched with glee as the mouse's fur began to burn. What we didn't realize was the mouse, being dead for several days, also smelled. Down in the basement we didn't smell a thing, but upstairs the smell permeated the room.

Soon our teacher, Mrs. Schroeder, hollered down the stairs, "What are you boys doing down there? It smells awful up here. You better quit playing around and come right up."

I quickly turned the frying mouse pan over and dumped the carcass into the burning coal. Any evidence of our wrongdoing went up in smoke. Jack and I scampered upstairs looking as innocent as possible. Mrs. Schroeder didn't trust us after that episode. Our cleaning the furnace time was pushed back to just before school dismissed. She figured we would stick to our job if we wanted to get home.

Winter Fun

SNOWY DAYS WERE good and bad. The snow which was damp enough to make snow into snowballs, was considered good. We made snow forts and snowmen. The snowmen had sticks for arms and walnuts for eyes. Their mouths were chunks of dark ash dug out of the ash pile.

Of course, snowball fights always were fun. We would construct two forts in the morning before school. At first recess, if it wasn't too cold, we'd mark out boundaries. Each opposing team could advance to the dividing line and no further. This made overrunning a fort against the rules.

Once in our snowball wars, Patty Hall and I were the only survivors of the west team. We huddled behind our snow fort wall while others pummeled us with snowballs.

I decided to start using our fort snow for making snowballs since we were dead ducks if we exited our confines. As Patty made the balls, I started to lob the projectiles like mortar shells. I got lucky and hit an enemy. It was Miss Kemper. She had to come to our side.

Now we had fire power. Soon we had another, then another. By the time noon recess was over, the west team numbered more players than the east team. We were attacking the east fort and about to declare victory when Miss Kemper declared a truce and said it was time to return to the classroom. I think she was having as much fun as her students.

That night it turned very cold. Our forts were solid ice, and the snow was impossible to pack.

We played inside for the next several days.

Good dry snow made areas good for the game of Fox and Goose. A big circle was tramped in the snow. Once the circle was made, we made pie-shaped paths splitting the circle into sections.

The rules were like the game of tag. One person was deemed the fox and the rest were the geese. The fox tried to catch one of the geese. The geese could only run one direction, but the fox could turn around. Everyone had to stay on the pie-shaped paths or on the outside circle.

We played for a while with eight sections, then someone would decide to expand the circle to two rings. Maybe make cutoffs and odd escape routes. It was a game all could play.

In the 1940s, plastic or waterproof clothing wasn't available. Our mittens were cotton or wool. Our coats and snow suits were mostly wool. The girls wore blue jeans or slacks under their dresses and skirts. Girls never wore jeans or slacks to school. All the clothing was made of material which got wet quickly from snow.

When the bell rang, we removed our outside clothes and hung them over the backs of chairs around the furnace grate. It was a great clothes dryer. During the class period, one of the older students would quietly go to the grate and turn our wet clothes over. There were times when the woolen mitten and gloves would become too hot and start to scorch. The smell of scorched wool is not pleasant. Our boots would also be placed by the grate. Then the smell of burning rubber blended with the scorched mittens.

The hill running across the northwest part of the school yard meant, depending on the snow cover, certain parts of the hill were better for sledding than others. The south hill was generally the best because it faced the north and held the snow cover longer. Its problem was the many gopher mounds on the surface made for a rough ride. The little hill north of the school building was steep and fast but short. What would country kids do? Search for bigger and better hills. Those slopes were just over the fence in Mr. Dipple's pasture.

Once there was a good snowstorm followed by an ice storm. The ice hung on trees and fences. It caused a lot of work for our fathers, but we saw it as a great opportunity to sled. The snowbanks were covered with a thick coating of ice strong enough to support a fast-moving sled. We could start on the big hill and skim over the ice almost to the little hill.

We had some rules. Once your sled stopped going forward, you couldn't get off. You had to slide backwards or sideways to the bottom. From there you could traipse back up for another run, walking next to the fence so as not to ruin the ice coating.

This was an experience one morning before school started that never happened again. By the noon recess, the sun had weakened the ice enough, the sled's runners cut through the cover of ice. It curtailed our sledding for a couple of days since below the ice the snow was deep. It was good only for toboggans, which none of us had

I remember another time after an ice storm, not only were the fields covered with ice, but the roads were also. The road which passed the school was a glaze of ice.

Our teacher had a very difficult time getting to school. She had to put steel chains on her car tires to grip the road.

Ice didn't bother us kids. It was an opportunity to seek some more thrills. Just south of school the county road sloped downhill for more than a quarter of a mile. We knew it was a one-day shot at excellent sledding, since traffic and a south facing slope would soon ruin the hill.

We all wanted to run that hill. We asked Miss Kemper if it was possible for us to go sledding on the hill. First she was reluctant, but being a farm girl herself, she relented. We had to make a deal because the hill was so long our recess would only allow one or two trips down. In order to receive the needed longer noon recess, we had to cut our morning and afternoon recesses to five minutes. This gave us a noon recess of eighty minutes. We were excited.

The morning passed slowly. At noon we gobbled our lunches and the girls pulled on their jeans or snow pants under their dresses. The boys sorted out the sleds and pulled them down the road. We stood at the top and marveled at the sheet of ice covering the road. This was going to be fast and long.

Miss Kemper and the girls followed a few minutes later. The fun began. Miss Kemper stood at the top and watched for cars. The recess was worth it. It was a long slide. The trip back up was a bummer. I think we might have made two or three trips down before it was time to go back to class. But what a day! That situation never happened again.

Sledding at Hazel Dell was a great sport. In fact, there were very few school yards in Muscatine County that had yards with a slope. Most of them were level. I would venture a guess that Hazel Dell had the school yard in the county with the greatest change in elevation. We had two possibilities for sledding outside of the schoolyard.

Ira Dipple owned the land north of school. His field had long steep hills. You could slide down and almost halfway up the hill on the other side. It was a kid's sledding heaven, but they were only available when Mr. Dipple raised hay on the slope. Sledding on cornstalks was impossible. We went sledding there only for one week before the weather warmed and melted the snow. It never happened again.

Henry Dipple owned the land surrounding the school to the south and west. Mr. Dipple's sledding hills were short, fast, and more accessible. Most of the bottom was flat. There was an area in which a ditch cut across the sledding lane. One could avoid it by steering away or stopping before hitting it.

One day our sledding adventures in Henry Dipple's pasture came to a screeching halt on a noon recess when Patty Hall had a sledding accident.

Patty was a little chubby. We kidded and teased her. In today's world, we'd have been reprimanded for bullying.

This day, we were coasting in the pasture. Patty was always timid when it came to sledding. Most of the boys rode our sleds belly-flop style. It gave you more speed. We'd pick up our sled and get a running start, then flop down on the sled. You could steer your sled with your hands. Most girls rode sitting up and steering with their feet on the crossbar.

Patty was always scared and dragged one foot in the snow. She never quite made it to the bottom of the hill. We expert downhillers teased and teased her. We called her chicken and scaredy cat. As she went down, we'd fly past her and laugh. "Don't drag your feet, Patty," we called to her.

Finally, Patty had had enough. She put both feet on the steering crossbar. I gave her a final push and off she went. As we cheered her on, we noticed she didn't know how to steer her sled. We could see from the top of the hill she was headed directly for the ditch on one side of the bottom of the hill.

Everyone started screaming at the top of their lungs. "Stop, Patty, stop." It was too late. Her sled careened off the edge of the embankment. We all stood stunned. None of us could see Patty.

All of us boys grabbed our sleds and roared down the hill, vocal sirens wailing. We slid to a stop just feet from where Patty was last seen. There, sitting at the bottom of the ditch, was Patty. Her nose was bleeding, and she was crying.

Jack and I jumped down to help her out of the mini chasm. Everyone headed for the schoolyard, except for me. I had to retrieve Patty's sled. It was stuck several inches into the dirt on the opposite side. I pried the sled loose and pushed it to the top of the ravine. By the time I climbed out, the rest of the gang had made it to the fence.

I called for help, but no one answered. The bell had rung by the time I reached the schoolhouse with the two sleds. Mrs. Schroeder was cleaning Patty's face. She glared at me as I entered.

"Bob, I hear you are the cause of this," she scolded.

"Me! I didn't do anything."

"Well, everyone said you pushed her."

"That's true, but I didn't cause her to drive into the ditch. I was still at the top of the hill. I was the first one down in the ditch to help Patty. Heck, I'm the only person who brought her sled back."

I glared at my schoolmates. They had turned on me like snakes just to keep their little noses clean. It was something which happened several times at Hazel Dell. Currently, I was the oldest student in school. Many times, Mrs. Schroeder asked for my help to tutor someone in the lower grades. I helped Patty with her arithmetic many times. I supposed the other students were jealous.

"Look, Mrs. Schroeder, I admit I gave her a push to start down the hill, but I wasn't the only one who teased her. All of us teased her. If I am guilty, then we are all guilty."

Mrs. Schroeder turned to Patty and asked, "Is that true? All the kids teased you?"

Patty nodded. Mrs. Schroeder thought for a moment and said, "Everyone to your seats. I'm going to call Patty's mother. I think Patty should go home. She has blood on her dress and her nose is very sore."

In a short while, Patty's dad arrived to pick her up because Patty's mom couldn't drive. Mrs. Schroeder let Patty's stepsister Barbara go home too.

Just before school was to be dismissed for the day, Mrs. Schroeder announced, "Tomorrow, there will be no more sledding outside of the schoolyard. If I catch anyone across the fence, you will stay after school."

We were dismayed. Our best hill was no longer available. This was our punishment. Mrs. Schroeder realized everyone was involved and not just me.

Patty returned to school the next day. She was okay. Nothing but a little scratch on her cheek. Patty easily dealt with a little pain. Farm kids are tough.

Snowball fights and sledding weren't the only outdoor entertainment. Iowa is noted for snowstorms with copious amounts of snow and high winds. The wind would pile the white fluffy stuff into deep snowbanks. The schoolyard seldom had deep banks, but the road ditches on the way home were filled.

The banks from the corner of the dead-end road where Jack and Joe lived to my lane yielded many tall banks. Sometimes the snow would be wind-packed, and a kid could walk on the top without sinking. I'd climb the bank and go a few feet before jumping on the edge and sliding down. An avalanche of snow followed. I would be covered with snow and half frozen by the time I made it home.

Mom greeted me with a stern look. She sent me to the basement to take off my wet snowy outer clothes. When I came upstairs, I was greeted by a warm cup of homemade cocoa and cookies. Mom wasn't angry after all. In a few minutes I thawed out. I still had my chicken chores to do.

Not all the days in an Iowa winter were suitable for outside fun. In fact, most days were too cold or too wet to go outside. I remember staying inside was never boring. We entertained ourselves with games like caroms, which had a slick board about three-foot square with pockets in the corners and a six-inch border painted around the edge. To start the game, we would place several red and green wooden rings in the middle of the board. Each player was given an uncolored wooden ring.

The goal was to shoot your color of rings into the pocket. You used your index finger and thumb to propel your shooter. With some skill you could move the rings into the pockets by caroming them from each other. The first player to get all their rings or caroms into the pockets won. If we didn't finish the game before the bell, we'd carefully lay it on a table. Occasionally, someone would bump into the board and cause the caroms to move.

Jacks was a very old game but still relevant for grade school children. We used a rubber ball and six jacks which were metal cross-formed pieces. The object was to throw the ball up in the air. While it was in the air, the player had to collect the jacks in different combinations before the ball bounced more than once. We became skilled at throwing the ball up and retrieving the jacks in the various combinations, like pigs in the pen, oneseys, twoseys, double bounce, and eggs in the basket.

The player who had the ball had to go each sequence. Once you missed gathering all the jacks in proper order, you were through with your turn. The person going doing the most combinations was declared the winner. Three or four players could compete or just two.

In the game of telephone line, we would sit on the long bench. The person at the end would whisper a message into the person's ear sitting next to them. They in turn would whisper what they heard to the next person, and so on. The last person in the line would tell everyone what she or he heard. It was always something different than it started out, and always a source of laughter.

In the game of Clap In, Clap Out, one student would be It. They would go out to the vestibule. While It was out of the room, the rest of us would give a hankie or rag to someone. They would sit on the rag. We would clap It in. He or she would walk around the room trying to figure out who was sitting on the rag. The room leader would say, "You're getting warm," or "You're getting cold." Finally, It would choose who was sitting on the rag. If It was right, he or she would replace the holder. If wrong, we would clap It out and start all over.

We did have games other than baseball when the weather was warm and dry. Andy, Andy Over was one where we would throw a softball over the schoolhouse. There were teams on both sides. If the team caught the ball when it appeared, they would give it to a player who would run around to the other side and try to tag those players. If tagged, that player was then part of the opposing team.

This would go back and forth until all the players were on one side or the bell rang. Some of the older boys could throw the ball hard enough and high enough to hit the roof just below the ridge. The ball would careen high into the air and make it difficult to catch. Most balls barely made it over the ridge and rolled down the other side.

One session, Gloria Kraft decided she wanted to try to throw the ball over the roof. She was only nine or ten years old. We called "Andy, Andy Over." She threw the ball. It went crashing through the schoolhouse window.

Of course, it scared the teacher. We ran inside to see the mess. Gloria started to cry. She figured she'd be punished.

Miss Kemper just looked at the bunch of us and said, "Well, I guess we will not play that game anymore. Somebody get the broom and dustpan. We must sweep up the broken glass."

No one was punished or scolded. Miss Kemper knew it was an accident.

To cover the broken pane, Miss Kemper cut a piece of cardboard the same size as the windowpane. One of us older boys stood on a table and taped it in. It worked until one of the dads repaired it.

Hide and seek was always a fun game, but one day, the game was intense. The entire schoolyard was in bounds, and just outside the fence. I was only a first grader, but I had discovered the perfect place to hide.

I don't know who It was, but he had to count to fifty so we could hide. I ran to the far southwest corner of the school yard. I crawled under the fence and under a brush pile on the opposite side. Under the pile was a washed-out space just big enough for me to scrunch down. I could hear each one of the others be discovered. I heard someone say, "The only one left is Bobby."

Soon the whole school was looking for me. Loren Braun climbed over the fence and walked right on top of the brush pile. I could see him looking down, but my brown jacket blended well with the dirt. No one could find me, not even Miss Kemper. Finally, she rang the bell.

I crawled out, but instead of going directly to the school building, I ran out of sight along the bottom of the hill and appeared on the north side of the yard.

Everyone asked, "Where were you?"

I never revealed my hiding place. I do remember sitting at my desk and combing particles of dirt from my hair. Loren watched me but didn't say a thing. The next time we played hide and seek, it had rained, and my hiding place was wet and muddy. The next year Mr. Dipple burned the brush pile, and my fabulous hiding place was gone forever.

About once a month, the school floor needed cleaning. First, one of the older students would sweep away the dirt with a large push broom. We dragged in a great deal of dirt and mud from our outside playground.

The teacher would open a large blue can containing a wax-coated red sawdust substance for the floor, called "sweeping compound." She'd fill the container inside and scatter the compound down the aisles and under the desks. The waxy particles were slick. We would run and slide on our shoes to see how far we could go. This was great fun.

The only problem was when some of the sweeping compound would fall into the furnace grate. The heating chamber below burned the waxy sawdust and caused a smokey odor. There was nothing we could do but let it burn away. Sometimes we had to open the windows to clear the air.

Luckily, this was only a once-a-month job. At the end of the day all the remaining sweeping compound was swept up and returned to the container and the floors were neatly polished. In time, the compound wasn't effective. It was disposed of, and new compound was applied.

Another activity one of the teachers presented was in writing class. We had writing class once a week for about fifteen minutes. The lower grades practiced printing while the upper grades did cursive exercises. One of the Kemper teachers decided to make the writing exercise more entertaining. First, she wrote out a letter on the black board to President Eisenhower. Everyone copied the letter and sent it to The White House c/o President Eisenhower.

Son of a gun, all of us received a signed photo of the president plus a large picture of him which we hung at the front of the room. We were a proud bunch of country bumpkins.

Two weeks later, we wrote to a celebrity we liked. I recall my choice was James Stewart. I believe all of us received a photo from the person we wrote. To rural children it was exciting.

Despite our active life in school, we still had free time. Jet planes were new. They were sleek and fast and used extensively during the Korean War. I was infatuated by the new airplanes.

One day my teacher gave me an old poster. It was plain on the reverse side. I drew spaceships, flying wings, and swept-back jets. When I finished, my teacher was somewhat impressed.

She said, "Why don't you send your drawings to the Department of the Air Force? Maybe they will be interested."

She was either pulling my leg or very naïve about the Air Force to suggest such a thing. Anyway, I agreed, folded up the poster, and mailed it to Washington, D.C. I figured I'd never get a response. The world fifty miles away from home seemed almost foreign to many of us.

But, lo and behold about four weeks later, teacher came to school with a letter from the United States Air Force. It read, "We received your many drawings. They were very interesting, but you must realize it takes many hours of work to design an airplane. Keep up the good work."

It was signed by some officer. I was as proud as a peacock. The United States Air Force had reviewed my ideas. I bet they all had a good chuckle in Washington. Heck, they probably hung it on their wall for an incentive.

Late winter days seemed to drag on. I longed for spring, which brought its own problems. The gravel roads became muddy, so riding bikes to school was out of the question. Everyone walked unless it was raining or snowing. On those days, parents drove us to school.

The road where Beverly Dipple and the Watts lived was not graveled. Their parents parked their cars on the corner at the end of the road and used horses or tractors to get to them. Beverly's dad transported her to school on a tractor. The Watts came in a horse and buggy. Horses seldom got stuck in the gooey mire of the Iowa mud. I remember wishing I could ride with Carol in the buggy.

By the time the Van Nices arrived in 1949, the road was upgraded and graveled. It was the last road in the township to be upgraded.

In late March, the snow started to melt. As the snowbanks melted, little rivulets of water trickled across the lower area of the schoolyard. The water came from Ira Dipple's field. It ran through a gap in the row of box elder trees which lined the north fence. Immediately on Ira's side was a ridge left from fieldwork the previous year. The water ponded upstream between the corn rows.

Running water attracts young farm boys like ants to a picnic. At noon the warm weather had melted the snow and the water was flowing. Inside the fence, the water ran over the fence lower board, where it formed a small waterfall on the schoolyard side. The cornstalks were deteriorating, and organic material mixed with the water.

As the stream tumbled over the edge, a dirty-looking foam formed brown soap suds. We tried to gather it and play in it, but the foam was very fragile. Someone took a stick and poked into the water to see how deep in was. It was over our four-buckle boots!

Jerry Yarrow was the oldest and biggest kid in school. Beverly Dipple and Marilyn Watts were the eight-graders, but Jerry was older. In today's world, Jerry would have been diagnosed with a learning disability. He was smart, but he didn't like school.

Anyway, he retook several grades. He lived with his mother on a small farm east of ours. I don't know if he had a father or not, since I never met him. Most of us picked on Jerry and teased him. I know he was an excellent artist. He had the ability to draw any object from a simple tree to an automobile with excellent detail.

This day, Jerry was with Loren Braun and me. The pool beneath this miniature waterfall intrigued us all. When we discovered the pool beneath the drop off was deeper than our boots, we looked at Jerry. Jerry always wore five-buckle boots. They were over sixteen inches tall. Surely, the water was not as deep as his boots.

"Jerry," Loren said, "I bet your boots are taller than the water is deep. Why don't you try and see if I'm right?"

At first Jerry said, "No."

Loren chided, "I dare you."

I added, "I double dare you." Those were fighting words.

Now Jerry had enough trouble making friends. Here was a time when he would be the hero. He approached the water hole. At first, he stepped on the edge. It wasn't so deep. He went in a little further. The water passed his third buckle. This was deep.

He placed his foot in the middle of the water pool. Down went his boot. Up came the water, third buckle, fourth buckle, and finally, the fifth buckle. The water was within an inch of the top of his boot.

He started to retreat, but his foot slipped. The cold icy snow melt poured over the top of his five-buckle boot and down inside. He immediately pulled away. The look on his face was one of shock.

Loren and I couldn't help but laugh. You see, when water gets inside your boot, it soaks your pantleg, sock, and shoe. Jerry could see nothing funny about it. He stomped away and headed for the schoolhouse. When the bell rang, we found Jerry sitting at his desk with a wet pantleg. He was sitting barefoot until his shoe and sock dried over the furnace grate. Of course, we said his sock smelled, just because it was Jerry's. He glared at Loren and me. He knew he had been tricked again.

Jerry was an okay guy. He'd push me on the merry-go-round. Once he helped me assemble a kite and showed me how attaching a tail made the kite catch the wind better. It was unfortunate that in the 1940s, there was no help for students with disabilities. They were simply considered "slow."

Jerry turned sixteen over the summer and by Iowa law was not required to attend school any longer. Jerry didn't return to school the next fall, and his family moved to Davenport. In later years, he ran a successful grocery store in the west end of Davenport.

Allen

NOT ALL THINGS at Hazel Dell were peachy keen. At the very end of the road where the Dipples and Wattses lived was a farm which was very isolated and rundown. The farm had been a rental property for years.

One year, a new family moved in. The family was a second marriage for the wife. She had two sons close to my age, Allen and Harold Paustian. She also had two sons by her present husband, but they were much younger.

Allen was a year younger than me and had red hair and freckles. If there was a Huck Finn in our neighborhood, it was Allen. He was easy going and lived with little cares, so we thought. His brother, Harold was a year older than me and quiet. School now had four boys: Allen in primary, me in first, Harold in second, and Loren Braun in sixth. Now, instead of Carol Watts, my new playmate was Allen.

Harold and Allen lived with an abusive stepfather. It was evident quite early that their stepdad did not like the boys. Maybe because he was dirt poor and feeding two growing boys was more than he could tolerate. Many times, the pair came to school dressed in hand-me-down clothes that were their stepdad's. Harold wore his stepdad's shoes with cotton stuffed in the toes to keep his feet from sliding out.

Another time Harold was late getting to school. Being tardy was reported on your report card. When the bell rang for school to start, someone saw Harold coming across the pasture. He arrived about five minutes late wearing knee-high boots and carrying his shoes. He told Miss Kemper he hadn't finished chores when his dad took Allen to school. He figured the shortest route was cross-country.

Miss Kemper accepted his story. I know she didn't count him as tardy because I saw his report card the next time it was issued. Miss Kemper had compassion for the brothers, but she didn't need to report these abuses. I believe other parents knew the situation, but since there was no actual evidence of abuse, they couldn't do anything.

The following summer, I would go to their place and play while my dad combined or baled hay for their stepdad. On one occasion, Harold showed me where his stepdad tied Allen to a tree by his ankle overnight for punishment. Another time Allen was locked in a grain bin for a day. It seemed Allen was the one always getting into trouble. Harold, being older, tolerated his stepdad's moods. Harold also could help with chores. I know he could milk cows.

The fall after Miss Kemper got married, we had the infamous Miss Geraldine Hidlebaugh. As mentioned, she was a poor teacher. She took her failures out on poor Allen.

One day Allen didn't have his work finished, or maybe he didn't understand the lesson. Allen sat in front of me, and Miss Hidlebaugh stood by his desk demanding his work. Allen ducked his head and tried to do what she asked. Soon Miss Hidlebaugh was whacking him across his shoulders with a foot-long ruler.

Everyone was stunned. We all had experienced minor punishment like sitting in the corner or staying after school, but this was violent. Allen just ducked his head and survived. I believe Miss Hidlebaugh finally gave up because it was either recess time or the end of the school day. We all thought, "Poor Allen."

My sister always felt sorry for Allen. One day Allen arrived at school with his face all scratched. He explained his stepdad had punished him by scrubbing his face with a steel brush. When my sister got home, she cried as she told Mom.

Today the school staff would report such abuse, but it was the policy in the 40s to let families solve their own problems. I am glad today there are family services personnel who can step in and help. I don't know what Allen's stepdad would have continued to do to Allen, but he may have abandoned him somewhere. Allen's mother seemed weak and helpless.

I remember Allen coming to school for a few days after the face scrubbing, then he was gone. In time my mother told me that his mother sent him to Annie Wittenmeyer Home for Orphans and Abused Children in Davenport. Harold reported a year later that Allen had been adopted.

I never saw Allen again, but I never forgot the incidents. Many years later, when I was an adult and farming the home farm, a friend of mine, Paul Barnes, phoned me. He wondered if I remembered a boy named Allen Paustian.

"Of course, I remember Allen. Why?"

Paul continued, "I received a call from a lady asking if I knew anyone who attended Hazel Dell School. I told her I knew just the guy, Bob Bancks."

Paul told me Allen was living in Prescott, Arizona, and his last name was Wilson. His adopted sister lived in Des Moines and was researching his early years. She had located Allen's brother Harold and the two brothers were meeting in the town where Harold lived. Paul told me I was invited but, unfortunately, I had a conflict and didn't get to visit Allen.

A few years later in an issue of the magazine *Our Iowa*, a story caught my eye. It was written by a lady from Des Moines. She wrote about the reunion of the brothers. Allen Wilson was a successful businessman, and the story described how his adopted sister did the research that reunited Allen with his brother, Harold Paustian. The story went on to describe their early days and the number of years they were apart.

I was enthralled by the story and wrote the magazine that I had attended school with both brothers. I hope Allen remembers good days and not the bad when he lived in our neighborhood. I am so glad he had a successful life.

Scammers

WHILE CITY SCHOOLS contained auditoriums or gymnasiums, country schools never had any school assemblies for entertainment.

One day a car pulled up outside the school and two men got out. Our teacher met the men at the door. They presented papers saying they were approved by the superintendent's office. I think it was why Miss Kemper agreed to let them in.

She returned and announced we had musical guests and they were going to perform for us. Classes would be suspended until tomorrow. The two men set up their instruments in the front of the room. One played a regular guitar and the other a steel guitar. To us students, they were very good. They sang and played for thirty minutes.

When they were finished, they asked if anyone would like to learn to play like they did. Of course, we all raised our hands. They took down everyone's name and phone number. They said someone would contact our parents to set up lessons.

Within a week Mom received a call from Bowlby Music Studios of Moline, Illinois. The visiting musicians were contracted by the music store to go to schools and sign students up for lessons and maybe instruments. They offered a six-week plan in which the company would provide instruction and instruments.

Lessons were held in an old building called the Central Turner Hall in downtown Davenport, not Moline. Mom was assured I would be tested, and an instrument would be issued. I had a choice of horns, woodwinds, or guitar. Mom thought I should try horn. She set up an appointment.

The following Saturday we journeyed to Davenport. The building was old and poorly kept. We were directed to an upstairs room. There we found several of our neighbors from Hazel Dell and the other schools in Montpelier Township.

Evidently, the township was targeted because of the closeness to Davenport. We all waited for our turn to meet the instructor. My name was called. We entered a little practice room and met a young lady. She was probably a grade school music teacher doing some outside work. She seemed very nice.

I tried a coronet. The instructor apologized for the poor instrument. She worked with me for thirty minutes and gave me some exercises and simple music.

The first thing Mom did when we got home was clean the mouthpiece. I was given the job of polishing the rest of the coronet. At first, I was excited. My cousin Lucy played in the Muscatine High School band. Maybe I could do the same.

Mom was okay with the project but still insisted I practice my piano. This really cut into my free time.

The next three weeks went okay. The only problem was, nearly every week, the instructor was different. We were moved around the dilapidated Turner Hall.

On the fifth week, Bowlby Music called and informed Mom the next lesson would cost ten dollars, plus a small rental fee for the coronet.

She claimed we had two lessons left for free. They explained their costs were more than expected and the

Turner Hall would be no longer be available. They had to rent a more expensive place.

By now, Mom was tired of the music company's wrangling. She told them she'd think about it. The next day she called some of the neighbors who had enrolled their children. Most had received the same call. Although the company was legit, they were not abiding by their agreement.

Mom called the store and said, "I believe you are not living up to your agreement. I found out you have changed the plans on many of my neighbors also. I am taking my son out of your program. I will return your undesirable instrument and I want my money back."

The person on the other end tried to talk Mom out of her decision, but in the end said, "You bring back the coronet and we will refund your rental fee."

The very next Saturday Mom and I drove to Moline to the Bowlby Store. When we entered, it looked like the usual music store with horns, guitars, drum sets, and counters filled with wood winds. A clerk appeared and asked if she could help. Mom handed her the coronet and explained her mission.

"I'll get the manager, Mr. Bowlby," she replied.

She disappeared and Mr. Bowlby came to meet us. He said he was sorry we were not satisfied with the program. He would try to make it up to us. He suggested Mom bring me to the studio in Moline for lessons. He blamed the poor conditions in Davenport and the poor instructors. Mom didn't buy it.

She told him, "I think you should have checked that out before you started, and I think preying on country schools to get business was not fair."

Mr. Bowlby could see he was losing the argument, so he answered, "I'll take back the instrument at no charge, and also give your son six thousand points to spend on toys from the Toy Cave, which is in the back of the store. Follow me."

We followed him to a little room filled with all kinds of goodies, baseball mitts, baseballs, toy cars, and my favorite, model airplanes. Six thousand points couldn't buy the baseball mitt, but it was enough for two model airplane kits. I made my choice. Mr. Bowlby graciously gave Mom her rental money back and we walked out.

Later we found out other neighbors had not fared so well. Some paid the extra fee for a while then quit. The Van Nices hung on for a few more weeks and even bought two guitars.

The Bowlby Store stayed open for many years. Mr. Bowlby was probably an honest man just trying to drum up more business, and that was my short experience of blowing on a horn.

Mary at the Bat

THE FIRST WEEK of school, we started a game of baseball on the lower diamond. Mary was one of the older students. Some of the girls in the lower grades had never swung a baseball bat or even played any kind of ball. We tried to tell them how to hold the bat and swing when the ball was on its way. It wasn't working, so Mary decided to show them the proper way to swing a bat.

This was the late 1940s. Girls were not supposed to be athletes. Girls did not swing from the shoulder and upper body like they do today. They held the bat waist high and swung with their hips. It was amazing how well a girl could hit. At least, that is what we boys thought.

Mary was at the plate, demonstrating how to swing. As she whipped the bat around, Sandra Jepson, a first grader, ran in front of her. Mary's bat caught Sandra across the bridge of her nose.

It made a crunching sound. Sandra stood stunned for a second, then her nose began to bleed. She turned and ran screaming up the hill. Miss Kemper, who was the umpire and playground teacher, chased after her.

Mary was in shock. She just stood there with the bat hanging from her hand.

Soon the rest of us collected the equipment and ran uphill too. When we reached the door, Miss Kemper had Sandra lying on the long bench inside the school. We all huddled around the injured girl. We had no school nurse, no telephone, and Sandra needed medical attention.

Miss Kemper said, "Everyone go to your seats. I must take Sandra home. I'll be back as soon as I can. Mary and Marjorie, you are in charge."

Miss Kemper helped Sandra to her car and left. Marjorie Westerhof took over. Mary was too shaken to do anything. Miss Kemper returned and reported Sandra was on her way to the hospital in Davenport. Sandra's mom didn't have car, so my mom drove Sandra and her mom to the hospital in Davenport.

Sandra was fortunate, nothing was broken. Three days later she returned to school with two black eyes and a nose with as many colors as a rainbow. The lesson we learned was to stay away from the plate when someone was batting. Years later, when Mary and I were going over our times at Hazel Dell, Mary said she still had dreams of hitting Sandra.

School Memories

HAZEL DELL'S LUNCH program was a lunch pail or Karo bucket stuffed with sandwiches, a cookie or piece of cake, maybe a carrot or slice of orange. My drink was carried in a thermos bottle. It was an insulated round canister with a hollow glass vial inside.

The glass vial was formed with two walls, between which was a vacuum. This vacuum kept hot drinks hot and cold drinks cold. The cannister was filled with milk, a majority of the time. Occasionally, in winter, Mom would give me hot chocolate.

There was a flaw in those thermos bottles. They were very fragile. Most of the students rode bicycles to school, even on cold days. The only days we didn't ride was when the road was muddy or snow packed. We carried our lunch pails in a basket mounted on the handlebars of our bikes. Because the road was not smooth, we tied our lunch pails in, but sometimes, a pail would escape and bounce out.

The fall jarred the thermos enough to break the glass inside. When lunch time came and you opened the thermos, out poured shards of glass in the liquid. For the next few days, our drink would be carried in a fruit jar wrapped in newspaper. It was a constant hazard until unbreakable plastic thermos bottles were manufactured, though I never had one.

The sandwiches Mom fixed for me ranged from peanut butter and jelly to egg salad, meat loaf, bologna, or boiled hot dogs. After Thanksgiving, I had turkey. After Easter, it was ham. One of my favorites was peanut butter and sweet pickles. I would go on binges and eat one kind of sandwich for several days. Many Mondays I had a chicken drumstick wrapped in waxed paper, a leftover from Sunday dinner. For dessert, maybe a cookie or two.

One week, Mom sliced Spam, a canned meat, fried it, and put it on bread. I liked the salty taste of the canned meat. It was a welcome change. I will say I had a variety of sandwiches. This was something all the moms had to do every day, fix a lunch for their children. There was no cafeteria or hot lunches at Hazel Dell. If for some reason you forgot your lunch, others would share theirs.

Another delicacy was cake, though cake wrapped in waxed paper seldom arrived in its original form. An apple or orange also in the lunch pail would squeeze the cake and cause the frosting to stick to the paper. I had to lick the paper after eating the contents.

Pudding was another dessert. It wasn't pudding from a box or a can, but real pudding made in a double boiler pan. Chocolate pudding was my favorite. The pudding was stored in a little glass jar, probably a used jelly jar. I ate with a spoon thoughtfully packed. To retrieve the last morsel, I'd stick my finger in as far as possible and scrape off the remaining goo. Licking my finger was especially good. Had I washed my hands before eating? Probably not, though we did have one teacher make us wash our hands before lunch. The only problem was water was difficult to get, we couldn't wash inside the school room, and we all washed in the same pan and water. Real good hygiene!

Our lunch hour depended on the teacher. When the weather was nice, most of our teachers allowed us to eat outside on the merry-go-round, on the hill behind the school, or maybe on the well platform. One teacher made us sit at our desks and eat. There was no conversation allowed.

She claimed someone had thrown their waxed paper away outside and this was our punishment. She really got on her high horse when some of us could eat faster than others or maybe didn't finish our meal because we wanted to have as much recess as possible.

Someone's mother complained about their child not eating properly because of the desire to go outside and play with the others. The rule then was, no one could go outside until everyone had finished their lunch. So, there we sat, watching the slowest eater calmly stuff their food away. He or she seemed to enjoy seeing the rest of us suffer. You know it had to be a girl. No boy would ever hold up play time.

I never put my lunch inside a refrigerator because the school had none. Our leftover chicken and egg salad sandwiches stayed in our lunch pail. It never spoiled or soured or got the dreaded e-coli or salmonella. I don't believe anyone ever got sick from consuming home-cooked food.

It's funny how different people have different values. Once our water supply was declared clean, the county health nurse decided the practice of setting our cups or glasses on a shelf was not sanitary. Instead of our own drinking vessels, we were supposed to use these little pointed paper cups.

The venture started out fine until we each used several cups a day. We were always out of cups. Soon our teacher decided we should all keep our cups at our desks and reuse them. Very sanitary, huh!

Every year our school put on a program or show. It was always an important event. We'd practice for weeks. The front of the school room became our stage. We strung four wires across the room, two from wall to wall and two tied between the first pair. Curtains hung from the wires to make a little stage.

The front curtain was a floral print. It was attached to the wire with drapery hooks so it could be slid open and shut. The other three curtains were pinned with straight pins and were made from a striped flannel material. Windows were paper and pinned onto the fabric.

Our only instrument was an old, upright, out-of-tune piano. It was set on stage outside the curtain. We always had some talented pianists play during the interlude between skits. The Watts girls, my sister, and I could play the piano. We all had our moment in the show.

It depended on the teacher when we performed during the year. Miss Mary Kemper preferred early in the year, like October or November. Her sisters Martha and Doris chose February, Miss Henke I believe was March, and Mrs. Schroeder always had the show the day before Christmas break. The programs were always well attended. All the parents came, plus many others from the neighborhood. It was a social night.

When I was a little kid, primary through second grade, I usually had small parts, maybe a poem or part of a group recitation. I did play a significant part in one skit as a first grader.

The skit was about someone eating a pumpkin pie. Loren Braun and Mary were the husband and wife. She had baked a pumpkin pie for Thanksgiving, and it was missing. She blamed her husband for eating the pie. He denied it. They argued back and forth while all the time I sat under the table eating the pie.

We practiced the skit many times, always pretending to have a pie and pretending to eat it. The night of the show, Miss Kemper brought a real pumpkin pie in a tiny pie pan. Just before the curtain opened for the skit, she smeared pumpkin pie all over my face. I hid under the table as was written in the skit. No one could see me because a long tablecloth covered my hiding place.

Mary and Loren repeated their lines. Everyone laughed. At the end, I crawled from under the table and confessed I had eaten the pie. I stood with my face covered with pumpkin. Mary and Loren were as shocked as anyone. The skit was a hit.

Most programs were uneventful, all of us trying to be quiet behind the curtain until it was our turn to shine. The teacher would announce every student's piece. On one occasion, Miss Kemper asked me to do a monologue recitation. It looked difficult. I refused and told her it was too much to remember.

She promptly asked my classmate, Jack. He accepted the role. Boy, was I embarrassed and angry with myself. I had given up the opportunity to be a star or at least have a prominent part. I quickly realized I had made a mistake. Jack, my classmate but also my friendly rival, was going to do my piece. The next morning before school started, I humbly asked if I could change my mind.

Miss Kemper said, "I have already given Jack the piece you refused. If you want to do a monologue, you must find a suitable one by yourself. Here are a couple of play books. You pick one out from there. I haven't time to find another."

She handed me the play books. I ate humble pie for the rest of the day. Jack never said a word, but he knew he would be the star for he didn't turn Miss Kemper down. I eventually found a suitable piece, but it wasn't the one I could have recited. It was a good lesson because I seldom refused a challenge again.

My last two years at Hazel Dell, our programs were just before Christmas break. Jack and I were the big kids. Carol Watts had graduated to high school. Harold Paustian had moved, so there was no eighth grade. Jack and I reigned for two years. During those two years, I directed the school programs. Mrs. Schroeder picked the material, but I oversaw helping the younger students learn their parts and the music.

I ran the music program because I was still taking piano lessons and sang in our church choir. Mrs. Schroeder had very little knowledge of reading a musical score. She was happy to have her eighth-grade student take over. It took some doing, but the musical part of the show happened. Some of my peers couldn't carry a tune in a bushel basket. Several years later, I learned some of my schoolmates were not appreciative of my efforts. They said I was bossy, and they were probably right, but we sang our songs with gusto.

Because I had my piano lesson in Davenport on Thursdays, Mom had an agreement with the teachers that I could be excused around 2:30. I didn't believe it was a privilege, but others did. One Thursday, we were going to decorate the school Christmas tree. In the morning, Mrs. Schroeder's husband delivered a nice fir tree. This was much better than our usual red cedar from Dipple's pasture.

The plan was to decorate the tree during fourth period. Fourth period was when I was excused for my lesson. I tried to talk my mom out of the lesson because I wanted to help decorate the tree.

Right after recess, we dug out all the decorations from the closet. We worked with the lights to fix the broken bulbs. The younger kids laid out the glass orbs and garland. It was 2:30. I looked out the window and could see my mother driving down the road. Man, I hated to leave. I knew what kind of a mess the others would put on our tree. But I suspect those left behind were glad to see me go.

Friday morning arrived. When I walked in the door, I could see the tree was a mess. Oh, the lights were okay, the glass ornaments were hung fine, but the aluminum icicles were just thrown on the branches. They lay in globs on the tree. I must have expressed my displeasure to Mrs. Schroeder. I started to remove the icicles and hang them properly. As each student arrived, I asked them to help straighten out the debacle. We had all the icicles hanging straight before the bell rang. Mrs. Schroeder had to admit the tree did look much better.

I didn't realize my bossiness irritated others until years later. My cousin, Sandra Jepson, and I were reminiscing once about our school days. She brought up how I made the others redecorate the tree and how I directed the music for the school program. I don't think she was upset but rather commenting on how Jack and I controlled the rest of school.

We were in charge of almost everything. Jack was the athletic director, and I handled the music and tutoring. Mrs. Schroeder allowed our bossiness because by this time there were twenty-four students spread throughout seven grades, and she was very busy.

Childhood diseases were common in Hazel Dell. If someone came down with the flu, it would spread all through the school. Chicken pox moved from one student to another. One would have a high fever and little pimples would pop-up all over your body. In two or three days the little pimples would form a head like a blister. These would erupt and drain. This was when the pox began to dry and the itch began.

Mom smeared calamine lotion on us. We would be covered with the white lotion from head to foot. As the little pimples dried, they left scabs. In the morning my bed was covered with little dots. Most times you would miss only a week of school, which in my case was meaningless. I was the only kid in the class. How could I get behind? My case of the pox was mild, but Mary became very sick. She missed many days of school. The dreaded pox formed inside her mouth and made it difficult for her to eat. Mom had her on a liquid diet for several days. Today children can be vaccinated and prevented from getting the disease.

Our mom wouldn't think of us missing a day of school except for chicken pox and other childhood diseases. I had to be almost dead before she wouldn't send me to school. If it was just a cold, she'd rub Vicks VapoRub on my chest and send me off. She never thought maybe I might infect others. If I didn't, someone else would, the disease just had to pass through the entire school.

Mumps invaded our school when I was in eighth grade. The younger students would be gone for a few days and be back. I was one of the last to be infected with the virus. Because I was older, it hit me hard. My glands swelled up and I looked like a pocket gopher. My throat was so sore I could barely swallow, though for some reason, Mounds candy bars slipped right down. I survived on them for a several days. I missed two full weeks of school.

One year, one of our Miss Kempers decided we should be clean and proper. She had probably attended some seminar over the summer. Her way of teaching cleanliness was we had to have our fingernails clean and a clean handkerchief each morning. This way we started out clean.

Each morning we would place our hankie on our desks and put our hands on the hankie. Miss Kemper would walk around and inspect us. If we had both items, we received a stick-on dot by our name. Of course, Carol Watts was always perfect. The rest of us tried to emulate her. I did well for a long time. Miss Kemper promised a prize for the student who received the most dots.

One morning, near the end of the school year, I was nearly to school when I realized I didn't have my clean hankie. I was devastated. Carol would win. I rationalized if I played sick and was taken home, my record would still be intact. By the time I walked in the door, I was sick. Miss Kemper asked me when it all started.

"Right before I got here. Oh, my tummy hurts," I moaned.

Instead of telephoning my mom, Miss Kemper took me home. Mom was surprised but didn't question my condition. She thanked Miss Kemper and made me lie down on the couch. The next day I was surprisingly much better. I walked to school since my bicycle was already there. I went through the daily inspection and passed.

The only problem was I never recovered my absent day. I had one blank spot on the chart. Carol won the prize handily, though I have no clue what it was. At least I hadn't been embarrassed by having no clean hankie.

Looking back on the event, I wonder why I didn't just turn around, ride home, get my precious hankie, and ride back to school. I had plenty of time. Oh well, Carol won even if she had only nine fingernails; she'd caught her finger in the cream separator when she was little and lost part of her finger. I lost the contest. It became another one of life's little lessons growing up.

At the time, these events seemed so important to me. I wanted to be the best. Maybe I wanted to impress my mother by being so responsible. Most of all, I wanted to beat Carol Watts.

During my last two years of school, during the reign of Mrs. Laura Schroeder, people from town were moving out to the country. Hazel Dell grew to twenty-four pupils. Mrs. Schroeder counted on her two older students, Jack and me, to help with chores.

When I was in primary the school mix began with three boys and nine girls, but now the tide had turned. The mix was eight girls and sixteen boys. Most of the boys lived on the dead-end road north of school. The other six boys came from two families. When school dismissed it was eight boys heading north, always ahead of the three or four girls who traveled the same way until the dead-end road, when they continued east. In fact, many times the girls' parents picked them up from school in a car.

One day we got to playing rough and started picking on David LaFrenz. He and his brother Kenneth lived at the end of the road. They were dirt poor. Kenny, as we called him, was picked up by his father because he was in primary. David rode his bike. Why we teased David, I don't remember. Today it would be called bullying.

We were about to the corner where there was a deepening of the ditch. I started to push David with my bike toward the edge. He tried to back away, but Jack parked his bike across the back wheel of David's bike. Eventually, with enough pushing, David and his bike tumbled into the ditch. He didn't cry, but he hollered, "I'll get you for this."

We all scrammed. The only problem was, at the corner, I rode east to my home while the rest of the gang rode down the dead-end road. David too had to travel down that road. Before he got home the dead-end boys had reconsidered their deed and talked to David. The next morning, as I was going to school, I met David's dad in his old beat-up pickup coming from school.

When I walked into the schoolroom, Mrs. Schroeder said, "Come here, Bob. I have a bone to pick with you." She explained. "Mr. LaFrenz was just here, and he is very angry about the event which happened yesterday. David said you pushed him into the ditch. Did you?"

I knew there was no reason to lie, so I answered, "Yes, I did, but I wasn't alone."

"I told Mr. Lafrenz I'd make you stay after school until David was well ahead of you."

I thought a minute and replied in my defense, "Now wait a minute. I admit I shoved David in the ditch, but Jack held his bike against his back tire so he couldn't move away. Joe threw his dinner pail in the ditch. I will take the punishment, but Jack and Joe must be with me. I take it the guys must have talked to David later to put all the blame on me."

Mrs. Schroeder went to the window and called Jack and Joe in. After further questioning, they admitted to being part of the event. To be fair, Mrs. Schroeder made the three of us stay after school for five minutes for one week or until the rest of the gang was far ahead and we had to promise to behave.

The next morning, I told David I was sorry for what happened. To be sure everything was okay, I chose him for my team the next time we played ball. As usual, kids forget and forgive. David and I soon were friends again.

Eighth Grade Exams

EIGHTH GRADE WAS a big deal in the rural schools. For many it was the end of a student's education, since in earlier years many children didn't go on to high school. By the time you graduated from eighth grade, you were fourteen years old. Your father needed you to work. There were many large families, and girls were needed to help in the home.

Sometimes, because corn was harvested by hand, older boys would not start school in the Fall until after the corn was husked. They would attend until Spring when there was planting to do. Most of those boys never actually graduated, for when they turned sixteen, they quit school. My dad never attended high school.

This practice led to the belief that many rural students might not be ready for high school. The Iowa Education Department made all students who wanted to enter high school take the dreaded eighth-grade exam. The city students took a similar exam, but most of them were destined for high school anyway.

In Muscatine County, all rural eighth graders went to the junior high school in Muscatine on a Saturday in April to take the test. We sat in classrooms with a desk between each student. One of the teachers gave instructions. We worked for about an hour and were given a break before another hour of exams. Boy, we were all sweating it out. I don't remember my score, but I passed.

Our next hurdle was orientation at Muscatine High School. All students met in a small theater. Seniors guided us around the huge building. At the end of the orientation, Mr. Clyde Gabriel asked us to pair up with a locker partner. I didn't want Jack and I'm sure he didn't want me. The other guys paired up, and I was being ignored by those I desired.

It came down to the guy sitting in front of me, Wayne Wetzel. I knew Wayne from 4-H. I tapped him on the shoulder and asked, "How about you and me?"

He answered, "Sure, why not?"

So, for the next four years, Wayne and I shared locker 232 with the combination of 15R-25L-5R. Now came the wait for next September.

The final event was the eighth-grade graduation ceremony at the Muscatine High School auditorium. It was a big event. The county choir sang, led by Mrs. Van Ysseldyk, a lady who was involved in the city school's music program. She was another skeptic of the rural school system. She couldn't believe all the students had her list of numbers memorized. Boy, did we surprise her. She praised us after our final practice and told us she wished her students in town were that good.

At the end of the ceremony, all the one-room country schools and their graduates were announced. We were proud when our school and our name was called. We dressed in our Sunday best. This was a big day because we were headed for the next level of our education.

The ceremony ended with a lady playing her xylophone. She always ended with a rousing rendition of "Star and Stripes Forever." Afterwards, we marched out and became high schoolers. I left Hazel Dell #3 behind.

The rule in the Shepard family was when you graduated from eighth grade, you received your first wristwatch. I remember going to Plank's Jewelry in Davenport to pick mine out. I think mine was a gold Bulova watch with an expansion band. Boy, was I proud! I only wore it to church and on special occasions. I was old enough to be trusted with a wristwatch.

As I looked back on my experience in Hazel Dell #3, I realized the perception that rural schools were inadequate was false. Hazel Dell produced two high school valedictorians and several National Honor Society students. Most rural students outclassed the city students when it came to academics. Very few excelled in sports, because our city counterparts had the advantage of junior high and gyms.

1955 was the last year for Hazel Dell #3. Consolidation was the word, and Hazel Dell was absorbed into the Blue Grass School District. For a few years, students were bussed around to updated one-room school buildings. Soon a new elementary school was built in Blue Grass. Hazel Dell was sold and moved to become a residence.

Jack Van Nice and I were the last to graduate from our wonderful neighborhood one-room Hazel Dell School #3. The rural schools became obsolete, and better roads and buses changed our world. Hazel Dell was built in 1866 and served my grandfather, my dad, and me as a great learning institution. Without those hundreds of one-room schools doting the landscape all over the Midwest, many of us would be very backward. They were the backbone of education and the social life of each community.

Sweetland Methodist Church

MY FAMILY ATTENDED Sweetland Methodist Church, a small country church located in a crossroads village named Sweetland Center. In earlier days, the interurban, a streetcar-like vehicle which ran on railroad tracks, stopped at the little town. There was a store with a gas pump, four or five houses, and the church. Iowa Highway 22 ran along the west side of the church. It was a graveled road leading from Blue Grass to Muscatine and beyond.

This was the road we took when the roads were muddy or icy. The route through Smokey Hollow was the quickest, but it was a dirt road and only available on dry Sundays. The third road cut through Wild Cat Den State Park. Many times, Dad took this route because at the New Era Store he could purchase the *Des Moines Register*, Sunday edition. If we were to be late for church, we'd pick the paper up going home from the Shepard's dinner. The owner always held a copy back for Dad. Regardless of which road we traveled, it was a twenty-minute ride.

In the early days there was a concrete platform across the front of the church. When horse and buggies were in vogue, this platform saved women the difficulty of stepping down from the buggy. During this same period there were covered horse stalls reserved for the wealthier givers. Mom told me when Mr. Grey, Grandpa Shepard's landlord, didn't attend, they could tie their team in his stall.

Before the first addition, the church was one big room with a center section of pews and two outside sections. It was heated by a huge grate above the furnace near the back of the sanctuary. Everyone entered through an unheated vestibule. Inside the building, there was a stage in front. The piano and choir sat stage left and the minister sat on the opposite side. The building had a basement for the kitchen and meetings. As in all early churches, there were no restroom facilities. You went to the outhouse. The church wasn't modernized until 1950.

The 1950s were God's years. Because of WWII and the Korean War, people turned to God for hope and grace. There was a surge in church growth and a surge in family size. Sweetland was no exception.

Sunday school classes were divided by portable partitions. The adults used various areas of the sanctuary. Each class used a set of pews for their class. The crowded situation led to the need for an addition to the building with modern restrooms. It would be a big undertaking for the congregation. The church building had not been changed for almost a hundred years.

Dad was on the board. The men tossed around many ideas. (Sorry, no women on the Board of Directors.) On which side should the main door be placed? Do we change the front windows? One man, Mr. Borgstedt, built a miniature building to express his thoughts. His model was used, except the entrance was moved to the north side instead of the south. It would be closer to the parking area.

Being mostly farmers, they planned on doing most of the work themselves. Dad contacted the carpenters who built our house to manage the job. By mid-June, the crops were cultivated, and the hay was made. It was time to start the project. My Uncle Jim told me that day several men were milling around the concrete platform, and none knew where to start. They all waited for Carl Bancks, who lived the furthest away.

The idea was to destroy the platform with farm tractor-mounted end loaders, my dad's and Gordon Day's. Gordon had to be there, but he was a cautious man and very careful of his tractors. As the men stood contemplating the first move, Carl came roaring down the road with his M Farmall and loader. He barely slowed as he approached the church platform. He and the tractor hit the concrete platform with a bang. The corner splintered off.

Carl stopped and laughed. "Well, boys, I guess we've started."

The ice was broken, and the church project began. Dad and Charlie Drumm were the chairmen of the building committee. Both men were pushers and leaders. By autumn, the church addition was finished.

Sweetland was my mother's church. If dad had one, he never spoke of it. My Aunt Georgie always claimed the Bancks were Lutheran. Mom never attended a different church, she never thought of going somewhere else. Her faith was deep. Dad liked the men, and all the Shepard family attended Sweetland. He just fit in. I remember when Dad joined the congregation officially. There was a class of youngsters also joining at the time. They all stood in the front and spoke their vows. Afterwards, the rest of the congregation paraded up and shook the new members' hands. I distinctly remember shaking my dad's hand. He later became a leader in the church.

On Sundays, I had to dig out my Sunday shoes, white shirt, and suit. We dressed up when we went to church, which was every Sunday. The only decision I made on Sunday mornings was what necktie I was going to wear. Not a big choice, since I only had two ties.

My chore Sunday morning was polishing everyone's shoes. Dad and Mom's just needed a touch up, Mary's maybe a little polish now and then. Mine needed polish every Sunday. I guess I was rough on shoes. There were several boys my age attending church, and after services in the Summer, we would play tag or catch on the big lawn. Invariably I would fall or slip on the grass, staining my good pants. Mother was never happy about my stains. I was always scolded, though never really punished. She rubbed on a cleaner from a little can called Energine, and the grass stain disappeared until next time.

We seldom missed church. Our vacations were planned around attendance. Perfect attendance was rewarded by a pin. The criteria to receive the award was you must attend fifty Sundays of the year. If you had missed your quota of Sundays, you could redeem yourself by attending a church or Sunday School elsewhere.

There were many who accomplished the one or two-year goal, but few beyond that. My cousins Jane and Jean Shepard, Janet Day, and I were a few who accomplished multiple-year pins. I may have made it to six years. Jane earned seven or more. Janet achieved the ten-plus.

Sunday school was after church services, depending on the minister. Sweetland shared a minister with another church. Some years he preached at our church first and other times at the other. Sunday school always ended with a gathering of all the people. We would sing some old gospel songs.

The Sunday school recorder read the attendance of each class and how much was in their collection envelope. I don't know how it happened, but maybe because of my mother's determination for me to have a good reason to attend, I became the Sunday school recorder and song leader. Cousin Jane was the Sunday school accompanist. I guess I was too young to become nervous. It was a good thing Jane was an excellent sight reader. I always had free will to sing anything I desired.

I remember those hymns. Beulah Kemper sat near the front. She loved "Wonderful Words of Life." She belted out the chorus with gusto. The men liked "Little Brown Church in the Dell" with its back chorus of "Come, come, come, come." I chose the hymns. "Old Rugged Cross" was Dad's favorite and "In the Garden" was Mom's. At the end of our sing-along, I would ask one of the men to give closing prayer. I never felt confident enough to do the prayer myself.

It amazed me how each individual prayed differently. I became accustomed to each man. Each had their own style. Uncle Vernon's were short and concise. Henry Sywassink's always ended with, "Bless the sick and afflicted." Gordon Day had his neighbors and the world in mind. Edgar Kemper's were simple and to the point.

David and Harlan Pace were the librarians. They were to distribute the class study lessons each quarter. Each Sunday they handed out the bulletins and my favorite, "David Cook's Sunday Pix." It was printed like a comic book with Bible stories and stories about missionaries. When Harlan and David graduated from high school, I assumed the job for two years. The last Sunday of each quarter, all the classes received new study guides for the next quarter. It required all Sunday school time just to sort and distribute the guides. Again, it was always boys or young men who ran the library. Girls weren't asked. I guess lifting the bundles of booklets was thought to be unladylike.

In September, one of the Sundays was designated as Rally Day. This was the start of a new Sunday school year. We had new classes, just like in grade school. There was a push to beat the attendance of 139 which was posted on a bulletin board at the front sanctuary. It had been the record for many years. Every year we would come close but wouldn't beat that record. The 139 stood out like a beacon.

The year after the remodeling, Gordon Day was superintendent of Sunday school. He announced his goal was to smash the attendance record. He started early with encouragements and letters. On Rally Day Sunday, the church bulged at its seams. Every class had a full room. I was the recorder at the time. Mr. Day and I added up the attendance twice to be sure. Yep, we hit 163 attendees. Mr. Day was elated. We couldn't believe it at first. The only time we had come close to having that many in Sunday School or even church was on Easter Sundays. I'm sure that number never was topped.

My first memory of Sunday school was of primary class. It was taught by a little old lady, Mrs. Pace. She would read us stories and have coloring projects. We met in the little room at the bottom of the stairs, boys and girls together.

Once we moved on from Mrs. Pace's class, girls and boys had tables separated by portable dividers until a person attended high school. Then boys and girls were together again. It seems strange now, but this was how it worked in the early 50s.

My other early teachers are a blur except for Gordon Day. He taught sixth to eighth-grade boys. By the time I reached that age the church building was renovated, but we still were short of classrooms. Mr. Day's fifth and sixth grade boys class met on the landing of the stairs in the new addition. There were three or four boys.

One Sunday, it was just Jack Van Nice and me. We read through our lesson quickly. The next task was collecting our offering. Each class had their own envelopes. Jack and I gave a quarter or two. We discovered if you held the coin on the edge and flicked it with your index finger, it would spin like a top. We spun our coins and giggled. Unknown to us, the noise from our coins was traveling up the stairwell to the men's class. All of a sudden, Charlie Sywassink appeared at the top of the stairs and said in a gruff voice, "If you boys can't put those coins in your pockets until class is over, I'll take them and put them in our collection."

It was the last time Jack and I spun our coins. Shortly thereafter, someone decided boys and girls could meet together in Sunday school class as we did in public school. I guess the split was a holdover from long ago when men and women were seated separately in services.

Every June we had Bible school. This was a time when we followed a theme and learned stories and teachings in the Bible. It lasted a full two weeks followed by a Bible school program. Anyone in the community was invited. It was open to children from four years old to twelve.

Every church in the area held one during the summer. I detested going. Elementary school had just finished in mid-May. I had playing to do, pastures and creeks to roam, but Mom insisted. She was one of the teachers, and once I got to Bible school, I enjoyed being with the other children. There were other Bible schools nearer to us in Blue Grass and Pleasant Prairie, but because Mom was a teacher, she dragged me off to Sweetland.

If it worked out timewise, I attended Blue Grass Presbyterian and Pleasant Presbyterian Bible Schools. Boy, I should have been educated on the Bible. Now I look back on those Bible schools and they taught me many bible stories and songs. I am sure I remember those stories much better than most children of today's world.

Besides the program after Bible school, there were two other church programs presented by the children. The first came during July and was called Children's Day. Instead of Sunday school, we recited poems, sang songs, and maybe the older kids put on a skit. It meant spending two or three Saturdays practicing our pieces. I complained to Mom once.

I said, "We give you a card and a gift on Mother's Day. We give Dad a necktie on Father's Day. Why do we kids have to put on a program for Children's Day? Why don't we get a gift or a day off?"

She glared at me but never answered.

The Christmas Eve program was the big production. Everyone who ever attended Sweetland Sunday school had a part. The best part was after the program, the church gave out goodie bags of apples or oranges, peanut brittle, wave candy, chocolate sugar mounds, and some nuts. Everyone received a bag.

Two programs stand out in my mind. I'd say I was six or seven. Nine children lined up across the stage. Each held a letter which when turned around spelled out C-H-R-I-S-T-M-A-S. We stretched from the pulpit to the piano.

I was at the beginning of the line with the letter C and stood next to the pulpit. I recited my three or four lines perfectly. Next was H. By the time we had reached S or T, I was bored. I stepped forward to see how far the letters were. Gosh, M didn't know her part. The mother in charge was giving her every word.

I fidgeted some more with my C. To Mom's dismay, I leaned on the pulpit and yawned. People in the audience smiled, but not Mom. She was mortified. Her little boy was making a spectacle of himself. It was a good thing we were early in the program because she calmed down before we went home. I was scolded, but there were no repercussions. After all, it was Christmas Eve.

One year, after the young children repeated their pieces, the older children presented the standard Christmas pageant. There was Mary and Joseph, the shepherds, angels, and finally, the three wise men. David and Harlan Pace and I were chosen to dress like wise men and sing "We Three Kings." This was my first solo. I sang the verse about my gift of gold in my high little boy's tenor voice. The three of us sang the chorus. I guess I did okay, since Mom didn't scold me like she did the time I was part of a line of poems.

At thirteen, I was considered old enough to officially join the church. There were several boys and girls at that age. We met with the minister on Sunday afternoons. I look back at that time now and feel sorry for Rev. Kennedy. He had to try and instruct an over-charged, anxious, peppy group of boys who were not interested in what he said. We were more interested in going outside and playing baseball in the church yard. There were some girls, but they didn't play ball. Those six or so Sundays were torture. Yes, we all accepted Jesus as our Savior at confirmation, but was it sincere. I think Jesus understood most of us would become good Christians when we grew to adulthood.

During the 50s, Sweetland Methodist Church flourished. Church was the thing to do and the place to go to meet your friends. The high school crowd was the most ambitious. On Sundays, there were never ballgames or other events scheduled in the morning. Sunday morning was reserved for church.

The high schoolers met in the kitchen. They were titled the Methodist Youth Fellowship or MYF. We sat around a big table with storage underneath for pies and supplies when the church women had a supper. It was a good way to pass notes back and forth. We were not very good students. Despite our teacher's effort, we were always into some mischief. Once when my Uncle Vernon tried teaching, the group was so disruptive he walked out. I guess he got their attention the next Sunday.

Just before our high school class went upstairs for the closing ceremonies, we planned the rest of our day. This was especially true of Mary's age group. In the winter, we would plan ice skating and sledding parties. There were plenty of hills in Sweetland or Montpelier Township for sledding. We'd ice skate on farm ponds. Many times, wind affected the smoothness of a farm pond. If it was windy during the freezing, the ice would have ripples on the surface. It called for strong ankles.

The Kemper family knew the manager at the State Fish Hatchery in Fairport. Many times, we'd use their ponds for our ice. They were big and generally smooth. Once we skated near the mouth of Pine Creek, which empties into the Mississippi River. It was near the Kemper's house. Their father, Edgar, checked the ice thickness in the morning. The party was planned during Sunday school class. Mary drove us down to the creek and parked along the railroad tracks.

Mr. Kemper had a fire made of driftwood going on the ice between the railroad bridge and the highway bridge. The ice was good because it was protected from the wind. Mr. Kemper lived by the river all his life, and he knew how it acted. He motioned for some of us boys to follow him. He started walking toward the river. We could see the river contained no ice. It was open water. That didn't bother Edgar. He walked right out to the edge of the ice where the Pine Creek emptied into the river. We boys hung back. This was no time to be brave.

Mr. Kemper was about two feet from the edge when he stopped. He turned and said, "The ice will be just as thick here as where you are skating. The river water cuts the ice like a saw."

He assured us the ice on the creek was very safe and over a foot thick. We skated under the bridges for a couple of hours. Afterwards, we went to the Kemper's home and had chili and hot chocolate. Beulah had nine children and was not intimidated by large groups of young people stopping in. She had a huge kettle of soup ready for us. Their kitchen table was large with a bench on the wall side instead of chairs. We were well fed and thawed out before we went home.

One cold January morning, our MYF decided to go sledding. It had snowed the night before. All the farm ponds were covered with snow, so skating was out. Rodney Brannen lived just behind a bluff on the Mississippi. His pasture had hills that were long and steep. The moon was to be at its fullest, so no flashlights or yard lights were needed. Rodney was also sweet on my cousin Jane and attended our church. He promised us his mom would have hot chocolate and soup ready after we sledded for a while.

We started sledding around six. It was cold, very cold. The temperature was close to zero degrees. We all dressed warmly. All, that is, but Rod. He refused to wear a hat to cover his ears. He said it was dorky. We all made two or three runs down the hill. Climbing up was a chore. We quit at eight, since we all had school in the morning. As we entered Rod's home, someone read the thermometer. It was six below zero.

When Rod's ears hit the warm room, they began to burn. His mom discovered he had frostbite on his ears. There was a scurry to put warm towels on his poor ears. In a few minutes, he was feeling better. As we ate our soup, we kidded the brave Rod. Earflaps may look stupid, but our ears were not frozen.

The MYF was composed of high schoolers until we had two new ministers. One was a student from seminary, Lowen Kruse. He was energetic. He and his co-preacher, Lester Moore, were preaching at three country churches and one small one in Muscatine. Pastor Lowen incorporated seventh and eighth graders into the group, and I became a MYFer. I could mix with my high school heroes without comments from my big sister.

Pastor Lowen also worked with the MYFs from Spangler, High Prairie, and Park Avenue Methodist churches. Since all four churches were yoked together, Reverend Kruse would have the four MYFs meet two or three times a year. This would encompass thirty to forty young people. All the MYFers from these churches went to Muscatine High School or would soon be attending. We all knew each other either from high school or 4-H. We had some great times together.

Pastor Lowen was almost too modern for the stiff-necked Methodist boards of the parishes. He proposed using Sweetland's basement for a square dance for the youth. It was the only basement with enough space for square dancing. All the boards refused to give him permission. Sweetland was especially upset. There would be no dancing in the Sweetland church. Rev. Kruse was not discouraged. He defied the board. He contacted a local farmer's organization, the Grange, and asked if we could use their hall. Luckily, there were enough non-Methodist members in the Grange to allow us in. We had our square dance. The church fathers were not happy, but it was time for a change.

While Rev. Kruse was at our church, the World Conference of Churches met in Soldiers Field in Chicago. He worked with his seminary in setting up the event. This was a monumental event. Ministers, priests, and other church leaders from around the world were there. Rev. Kruse thought our MYF group should attend if he could get tickets. It was a once in a lifetime event. He would be one of the participants in the opening ceremony.

The four churches combined to fill a bus. We got permission from the church board to miss church that Sunday. It was a summer event, so going to school the next day was not an issue. Soldiers Field was packed. The place sat over 100,000 at that time. We tried to pick Rev. Kruse out from the many people on the field, but we were too far away. The most memorable event during the ceremonies was when a hundred thousand people stood and sang "Amazing Grace." When it was over, no one slept. We sang all the way home. We kept the driver very awake.

Unfortunately, the next year Rev. Kruse was transferred to a church nearer to his home in Nebraska. It was and still is the practice of the Methodist Church to move their ministers every few years. MYF survived, but it never was as prominent again. Sundays became another weekend day when stores opened on Sunday. Soon youth baseball and other sports crowded in. Church and church youth groups became secondary.

One task for the young men at Sweetland was to light the candles and turn on a painting of the Last Supper above the altar, plus assist in taking up the morning offering. We sat in a shortened pew at the back of the sanctuary. Because of the characters using this pew, it was dubbed Sinner's Row. When offertory time was almost upon us, we would point at each other to decide who was up this Sunday. Some of the teens refused, some would do it once in a while, and guys like David and Harlan Pace and me, who didn't mind the job, worked every Sunday.

I will say we were not little angels in the back pew. Some days we would receive the stare from the adult men or one of our mothers. One minister, Reverend Stewart, was long-winded. He was an older gentleman and near retirement. His prayers went on and on. It got so bad the guys in Sinner's Row would time them. Each of us would guess the length right down to the second. One member of our group was designated the clock man. There were no prizes, just a thumbs up to the winner.

The pastor was so long-winded with his sermons that his wife would sit near the back. She knew he had another church to serve that morning, so she watched the clock. If his sermon was dragging, she would stand up and wave. We in Sinner's Row would giggle. Reverend Stewart was the only minister the elders asked to have replaced. He lasted about one year.

Unfortunately, like many rural churches Sweetland has declined. There just aren't that many farm families any more to support them. For a while members would drive out from Muscatine, but they too grew older. The young families had children involved with sports and other events. Soon only older members came to services. The church still exists, but very few still attend. Sweetland Methodist Church will always have a special place in my heart. It was where I grew up with all my friends and cousins. We were one big family looking after each other.

Almost everyone attended a young woman's wedding. The reception was held in the church basement and catered by the women's group. There was no big dinner and dance at an event center. Weddings were simple.

If a funeral was held there, it would be well attended. It would be followed by a big lunch in the church basement after the interment. It was a time to visit with folks you seldom had contact with or who maybe came a long distance. Now, because of expense of visitations and lack of interest, funerals are short with no visitation. We have become a society too busy to lend a hand to a bereaved spouse or family member.

Miss Montpelier

IN THE EARLY 1900s, farming was just beginning to become mechanized with various labor-saving machines. Hybrid corn and fertilizers were being introduced to agriculture. Land grant colleges were established and from them the county extension offices grew. Their goal was to become an extension of the college or university.

The extension agents held meetings and demonstrations throughout the year. Many of these meetings were held in the winter months when farmers had more time. The meetings were tagged by the extension personnel as the farm bureau of the extension department. Soon they organized into regular meetings and the organization now called Farm Bureau was born in the years between 1914 and 1917. From then on until the late 50s, county, state, and national Farm Bureaus flourished.

Today, Farm Bureau is still active, but their perspective has changed. Local township groups are no more. Farm Bureau has many large co-operative ventures and have a large lobbying group in state legislatures and U.S. Congress.

In the 1940's Montpelier Township conducted its meetings in the schools. Not only did the extension agents attend and speak on different agricultural subjects, but the meetings were also a place to gather and exchange information and neighborhood gossip. They were always well attended.

It was a great time for the boys. Patterson School had many boys who later became great friends through 4-H.

As our parents visited and conducted a meeting inside, we played outside if the weather permitted. After the adult meeting was lunch. Generally, the lunch was cake and cookies. The adults had coffee and the kids had scorched cocoa. Invariably, the ladies would forget about the cocoa warming on the tiny hot plate each school had. None of the schools had running water, so the scorched cocoa was all we had. Rural frugality would not allow for a pot of cocoa drink to be thrown out. After a few sips, you got used to the burnt chocolate taste.

The most important meeting for the kids was the annual Halloween costume competition in October. Because the meeting was held on the third Thursday of the month, our Halloween was never on Halloween. Nevertheless, we would dress up and parade around the school room in many different costumes. Scarecrows, princesses, and hobos were the most popular. We weren't fancy, just clothes and rags gathered from Mom's throwaway closet.

The year I was eleven, I was in a quandary. What would my costume be this year? I had been just about everything I could conjure up with the little resources we had. The meeting was on Thursday, and it was already Saturday. Heaven forbid, if we would go to town and buy a costume.

Mary came up with a brilliant idea. "Why don't you dress up as a girl?"

Everything seemed to click. We found some old dresses of my sister's and her old shoes. The only problem was my hair. How would we disguise my hair? We couldn't afford a wig of any kind. Mary and I thought and thought.

We were about to revert to the hobo routine when we turned on the old Capehart black-and-white television that evening. There was a movie which featured Esther Williams, a movie star and female swimmer who held many records. She and other actresses in swimsuits would splash around in a pool. Nothing novel about that except the women all wore bathing caps. Bingo! I could go as a bathing beauty and cover my hair with a bathing cap.

Miss Montpelier was born. We found Mom's old wool bathing suit and a bathing cap of Mary's. My boyish body fit the suit perfectly. Mary taught me how to walk in some of Mom's old high-heeled shoes.

On the night of the big show, Mary and Mom helped me dress. Luckily, the Farm Bureau meeting was at Hazel Dell, which was close. I could dress at home. I squeezed into Mom's suit. Mary tucked my hair under the bathing cap. We made up a cloth banner with the title of Miss Montpelier printed on it. Mom safety-pinned it across my chest. She found some old lipstick, rouge, and powder. Both Mom and Mary helped apply the makeup. I decided I wouldn't put the high-heeled shoes on until right before the parade. At the last minute, Mary stuffed two hankies in the suit to give me little school-girl breasts.

Since we could see the school from our front window, we waited until several cars had arrived and the meeting was about to start. When we got to the school, I stayed in the car until Mary came and got me. We didn't want any of the others to see me early.

The quarters collected from the parents were divided into piles for each category. This was the prize money. The parade started. The judges sat in the front of the room. I put on my sweetest smile and flashed my eyes. I was the last in the parade. I carried myself with such pride. I never faltered in those shoes. Everyone was guessing, who could the little girl in the bathing suit be, because there were few girls in the community of this age. After three circuits around the room, we were asked to line up in the front of the room.

There were about ten of us. The judges started to give out the prizes. The best hobo, the scariest monster, best scarecrow, etc. They had categories for everyone. Finally, they got to most beautiful. Mr. Kemper, one of the judges, said, "And the most beautiful girl and top prize goes to Miss Montpelier."

Then he said, "You can take your mask off now, little girl."

Boy, did I surprise him. I took off my bathing cap to reveal I was not a girl at all. I had fooled everyone. The crowd laughed. It was the best costume I ever had, and I will never forget the night I was crowned Miss Montpelier. Plus, I received four quarters for top prize.

The Cousins

THE SHEPARDS, MY MOM'S family, had dinner together almost every Sunday after church. In rural Iowa, dinners were served at noon and suppers in the evening. The family included Grandma and Grandpa Shepard, Uncle Vernon and Aunt Florence, Lucy and Sue, their daughters, Uncle Jim and Aunt Harriet, Jane and Jean, their daughters, and the Bancks family.

Once in a great while Uncle Arlo and Aunt Helen would show up with Bill and Anne. They lived in Letts, and it was quite a drive for them. All the children became known as The Cousins. We played and ate together.

The meals were held at one of the family's homes. We rotated homes each Sunday, but we ate at Grandma Shepard's more than the others. During the week, Grandma would write on a penny postcard what each of the women were to provide. The card would say, "Edna you bring dessert, Harriet, vegetable, Florence salad, and I'll furnish the meat and potatoes."

The meat was generally one of two things, fried chicken or meat loaf, furnished by the host. Occasionally we had ham. They were products we raised. Potatoes were always mashed and not just boiled like we had at home.

Each mother had her special dish. Aunt Harriet's was a dessert we called Fizz. It was pineapple mixed with thick whipped cream on a bed of crushed graham crackers. Delicious! Aunt Florence made many good items, but her divinity candy at Christmas time would melt in your mouth. Mom's specialty was a chocolate cake made with a boiled fudge frosting. The frosting would harden when it cooled. Grandma Shepard's dish was escalloped oysters at Christmas. The oysters swam in an ocean sauce of rich Jersey cow cream. Grandma was also known for her coconut cream pie.

After we ate our meal, Jane and I were allowed to change into play clothes. Jane seldom became dirty. It was me they worried about. The women washed the dishes while the men visited. Around four, we headed home for chores. Everyone milked cows.

Before going home, the leftovers were divided. The next day my school lunch would be a chicken leg or meatloaf sandwich. If Mom made cake, a piece of that was wrapped in waxed paper.

There were Sundays the men knew from church conversation about someone with a new piece of machinery or new building. Maybe one of the uncles had some new cows or purchased another forty acres. They would leave after Uncle Jim had his nap before they drove to inspect the new venture. This left the cousins to entertain themselves.

The cousins were divided into three groups. Lucy, Sue, and Mary were the eldest, the big three. They were less than two years apart in age, but four years older than Jane, Bill, and I, the little three. We were less than a year apart in age. Jean and Anne were the youngest. They didn't participate in our cousin adventures since they were considered babies.

Because Bill and Anne lived in Letts, our gang usually was just me and four girls. Lucy was the leader of the pack. She was the eldest and had a vivid imagination for play. Some Sundays we would be pioneers. We rode our make-believe covered wagons and fought the bad guys. When Lucy ran out of something for me to do, she'd send me out to hunt a bear. Jane was always a make-believe sick girl so Mary and Sue could nurse her back to health.

In the summer, Lucy, Sue, and Mary would make hollyhock dolls. They held weddings, and dances with the princesses of Hollyhock. One Sunday in late March, we were at Sue and Lucy's. They had acquired a large wooden button from who knows where. It was a nice day in early spring. They asked permission to go to their creek and float the button.

We walked north of their house to a creek. I was excited that I was going to explore a new waterway. The creek was big enough to float the button, but it was nothing like our creek at home. Why, you could step across it. Nevertheless, we killed an afternoon chasing the button downstream.

Lucy and Sue also had a horse. When there was enough snow Uncle V, as we called him, would hitch the horse to an old sleigh. We'd take rides out in a snow-covered hay field or pasture.

To me the best times were when we had our dinners in March at Grandpa Shepard's. Grandpa sold Pioneer Hybrid Seed Corn. He was one of the first men to do so. In early March, the seed corn company delivered the seed Grandpa had sold to the local farmers. He'd stack it in piles in his basement according to the size of seed and variety of corn. There were different seeds for different purposes. For three to four Sundays, until he had delivered the corn to his customers, Lucy would dream up great tales using the layout of bags of corn. They were stacked in the small piles and provided narrow alleys. We had to be careful we didn't cause one of the bags to slip from the pile. If we did, we'd work to put it back.

It is a wonder we didn't become ill from playing on the bags of corn. All seed corn was treated with a chemical called Deldrin. It was an insecticide to prevent insects from damaging the seed after it was planted. At times our rambunctious playing raised the dust from inside the bags, which were cloth and not sealed plastic and paper. We coughed and sneezed some, but we never slowed down. Deldrin was banned as a seed insecticide several years later. We were probably lucky we didn't contract a lung disease.

One memorable photo captures the cousins perfectly. The background tells me it was probably Mother's Day. The trees in the background have very few leaves. Bill was in the photo, so I assume they drove from Letts for the Mother's Day celebration. The photo is of the six cousins. Jane, Bill, and I are wearing toy six guns. There must have been some bad guys at our house that day. Lucy stands by Sue, who is holding a large limb at one end while Mary holds the other end. In between, draped over the limb, is a dead groundhog.

A groundhog is an unwanted varmint on any Iowa farm. Just how did this groundhog get himself in this predicament? This is the tale of the great groundhog caper.

On this warm and sunny Mother's Day, the Shepard and the Bancks families went to church. It was Shepard policy not to do work on Sunday. They did chores such as milking, but no field work. This Sunday's meal was held at our place. It was the usual fried chicken and roast beef, mashed potatoes, vegetable, and pie.

The men retired to the front room and the women finished dishes. The cousins were allowed to change into play clothes. The big three changed from Sunday dresses to everyday dresses, no blue jeans or shorts. Bill and I got into our play clothes while Jane changed into her pedal pushers. Jane was allowed to wear pedal pushers or capris way before the older girls.

Lucy promised an exciting afternoon of pioneer living. This was going to be good since we asked permission to hike and play in the adjoining pasture, which had many rolling hills and sharp ravines. We walked through the dairy barn to the pasture. It was down a little hill, across a small dam. On the other side was an ancient cottonwood with its roots exposed.

Lucy sat on one of the enormous roots and explained the story. Jane, Bill, and I would go ahead with our trusty six shooters and clear out the bad guys just over the hill. Lucy, Sue, and Mary would follow shortly and set a claim on the land. The ploy really was to get rid of the younger set so the elder girls could talk. They were becoming young ladies and these childish games no longer had the same appeal.

The three gun carrying cousins followed the trail through the oak trees and over the hill. We were shooting bad guys and Indians left and right. To our surprise, and I'm sure to the groundhog's surprise, we discovered a fat groundhog out of its den lolling in the sun.

The groundhog ran along the top of the hill hoping to circumvent the intruders and return to its den on the other side of the gully. All three of us began to yell and holler at the varmint. Hearing our screaming, the elder three came running. After all they were supposed to be "babysitting" us.

Once over the rise, they realized the situation. Lucy took charge as always. She ordered Sue and Mary down in the ravine. Groundhog, now referred to as Hog, made a nervous mistake. Instead of continuing around the upper end of the gully and to his den, he decided to charge across the gully directly. As he tumbled down the east side, he was met by Mary with a big limb. She batted at the terrified animal.

He retreated and lumbered up the ravine only to find Sue waving another limb, which came mashing down, hitting him along his head. Hog made an about-face and spied his sanctuary only twenty feet up the steep bank. He made a dash for it. Lucy called in the reserves. Jane patrolled the east side while Bill and I were stationed on the west side where the den was located.

As soon as Hog started up the embankment, Lucy yelled, "Get a rock, Bob. Bill, find a stick." Bill threw a stick at him. I found a small rock and threw as hard as I could. Hog retreated and was met by an amazon named Mary Ann. She thumped Hog on his head. He staggered and turned.

Lucy called, "Watch it, Sue! He's coming your way."

Just as Sue was about to take a gargantuan swing, Hog started up the side patrolled by Jane and General Lucy.

"Help me throw this branch, Jane," Lucy ordered.

The two of them heaved a four-foot branch which Jane had dragged down from the grove of oaks. Hog retreated again. This time Sue didn't miss. She brought her weapon directly down on Hog's head.

Mary ran up and pummeled him with her walnut staff. The two amazons beat poor Hog until blood appeared from his nose. He lay motionless. Sue prodded him with her branch. Hog seemed dead.

Mary exclaimed, "I think he's still breathing."

She took another whack. She and Sue waited another minute.

Brave General Lucy called down from her post high on the bank. "Poke him with your stick, Mary Ann."

Mary poked and turned Hog over. He was limp.

"I think he is dead," she called up to Lucy.

"I'm coming down to make sure," said Lucy. "You little kids, stay up on the bank."

The General made her way down the embankment. She bent over for a closer look at the now deceased Hog.

"Yep, I do believe he is dead."

The three reserves scrambled down the bank. We stood there amazed at what we had accomplished.

"What do we do with him now?" asked Sue.

"Yeah, what now? Our parents will never believe us," added Mary.

"Let's haul him to the house," said I.

"I'm not touching that thing," announced Sue.

"I know. Let's carry him over a log like the hunters used to do," said Mary Ann.

"Excellent idea," confirmed Lucy.

We grabbed one of the straighter and stronger branches. Somehow, we pushed the groundhog onto the branch. No one would touch the beast. With much effort, Mary Ann and Sue climbed out of the ravine and did not drop the limp Hog. It was three or four hundred yards to the house.

The reserves went ahead to announce the conquest. Of course, all the parents and grandparents came to see. Mom snapped a photo to record the great victory.

I don't know what happened after that. I'm sure we all had to wash our hands before going home. Probably we were fed a piece of cake or cookie and drink. Dad probably disposed of the poor animal later that evening. Anyway, the cousins had rid the landscape of one varmint all by themselves.

We cousins spent many Sunday afternoons together. Once Lucy, Sue, and Mary entered Muscatine High School, Sunday afternoons for them involved studying and trips to Muscatine. Soon it was boyfriends and girlfriends. Bill moved to Texas, so it was Jane and I who played our Sundays out.

Jane's sister, Jean, who was three or four, was too young for our imaginative minds. There was a period of three years where Jane and I were together many hours. Not only on the Sunday afternoons, but during the summer we stayed at the other's homes for several days at a time.

Jane's birthday of July 19th coordinated with an annual event on the Muscatine levee which featured a carnival plus a 4-H beef show. To celebrate Jane's birthday, Uncle Jim and Aunt Harriet would take us there for the carnival rides. Aunt Harriet let Jane and me ride the Merry-Go-Round and Ferris Wheel. I guess Jean was along somewhere. She probably rode the kiddie rides.

Uncle Jim liked the beef show. The judging was not at night, but many of his farmer friends were there and Uncle Jim loved to visit. On the way home we would stop at Wintermute's Ice Cream Shop for a cone. It was a glorious ending to a warm evening.

The summer was the best time for Jane to come for a visit. Uncle Jim would be hauling a load of hogs to the packing plant in Davenport, and he would drop Jane off at my place on the way home. She would stay for three or four days.

Jane wore shorts and pedal pusher pants most of the time she came to visit. But one day, she arrived in a dress. Uncle Jim and Dad were going somewhere together, so she rode along to my house. It would not be an overnight stay. We played behind the old barn where Dad had pastured some hogs. There were big hog wallows. The rainwater from the barn roof supplied the water. It was a soupy mess.

Being farm kids and not afraid of most animals, we climbed over the gate into the pasture. Our intentions were to venture to the little pond behind the next hill. One of us picked up a stone or clod of dirt and threw it into the muddy pool. Hog wallows don't splash like water. They throw up sprays of mud in little droplets. The mud doesn't close over quickly; a clod of dirt leaves a depression like a small meteor had hit. This was fun. Jane and I tossed bigger and bigger projectiles into the goo.

We heard my mom call, "Jane, Bobby, where are you?"

I called back, "Behind the barn."

Soon she looked over the gate at us. She gasped.

"What are you two doing? Look at you."

We then realized the smelly, slick mud from the hog wallow had stuck to our clothes in many places. Jane's dress was a total mess. Another problem we found was that mud from a hog wallow stains cotton fabric.

Mom was not happy with us. I was lucky Jane was there or otherwise my butt would have been sore that night. Mom herded us to the house, took us to the basement, and stripped both of us to our underwear. Once upstairs, she dressed Jane in one of my overalls and a shirt. We almost looked like twins except for Jane's longer hair.

Cleaning Jane's dress was not as easy as it would be today. Mom did not have an automatic washing machine. Her washing machine was a tub Maytag which she had to fill with water and soap. Instead, she washed, or tried to wash, Jane's dress by hand. Of course, Dad and Uncle Jim returned before her dress was dry. Jane wore my clothes home and Mom sent her damp dress home in a bag with a note apologizing for the mess.

Other times when Jane visited during the summer, our favorite pastime was playing in our creek in the pasture. The small creek Jane had at her home was fed by tile lines and many times was barely flowing or dry. Our creek was huge compared to hers. It had many twists and turns. It generally was two to three feet wide and maybe knee-deep to a child in places. In many places it had a sand bottom, or almost sand. When the creek made a sharp bend, it cut out a deep enough hole for minnows and tadpoles. In the summer, our main purpose was wading to try to catch minnows and frogs.

When we were six, seven, or eight, Mom would let Jane and I go to the creek by ourselves. She knew we would get wet, so she dressed us in a minimum of clothes. Jane wore her shorts, top and underpants. I wore my overalls and underpants. I was a boy and needed no shirt. Both of us were barefoot. Mom or my sister would fix us a small lunch of a couple of cookies and a jar of water. Neither of us had a watch, so Mom packed one of Dad's old pocket watches with the cookies.

It took a good twenty minutes to reach the waterway. The day was always hot. We started at the tractor crossing. It was shallow and sandy. Downstream at the first bend was a nice sandbar. We'd float our pieces of bark or hickory nut shells. Soon I was soaked to my knees. The bottom cuff of Jane's shorts would be wet. Eventually, one of us would slip or intentionally sit down in the cool water. We'd splash each other until we were soaked. Our clothes became heavy and cool.

I don't know who suggested it—probably me—but one of us said, "Let's take our clothes off. We have underwear on."

The other agreed and quickly we were splashing in our underpants. Now at this stage of our young lives, boys and girls were not much different physically except below the waist. So why shouldn't we have fun?

We decided to try skinny dipping although we never heard of the term. It just seemed natural. There was no one in the pasture but some Holstein cows. The nearest house was a good quarter of a mile away. Off came the underwear and we stood *au natural*. I am assuming childhood curiosity affected both of us. Jane only had a little sister, and I had an older sister. I remember scanning each other for a few moments, commenting on how we were different, end of discussion. The rest of our creek time we spent without clothes.

When it came time to return to the house, we put on our damp clothes and trekked up the hill. By the time we reached the house, our clothes were almost dry. We changed into some dry duds and continued our day into evening. This didn't happen just once, but several times.

I think we knew we were not supposed to go skinny dipping, or our parents would scold. Somewhere or somehow our mothers had told us this was taboo, but it was fun and daring. It was much easier to wash off a smudge of mud from a bare butt than trying to do the same from cotton underwear.

At times I would stay for a day or two at Jane's. Her father had an old barn with a haymow which went from the hay track to the ground, with bales of hay stacked at various heights. It was a great place to climb and play with the many kittens living there.

Jane lived in a big square farmhouse with four bedrooms upstairs. They only used three of them. The fourth was a storage room. Jane and Jean each had their own room. Her parents had one large room, but there was no guest bedroom. Instead of Jane sleeping with Jean while I was visiting, I slept with Jane in her big bed. We were only eight or nine years old, and we wore pajamas.

Grandma and Grandpa Shepard lived across the road from Jane. We'd venture over to their house often. Grandma always had cookies and milk. Jane and Jean stayed there many times.

We were playing house in the upstairs bedrooms when Jane showed me a secret hiding place. In a walk-through closet upstairs between the two rooms, Jane showed me a loose board. Underneath the board my grandparents kept their valuables and important papers. We never dared to touch anything. I figure after living through the Great Depression and losing their farm, they didn't trust banks anymore. Grandma also had a very lumpy carpet in the living room. Underneath it she stashed many other documents and letters she deemed important. Everyone in the family knew what was under the lumpy rug.

Grandpa slept in a separate bed in a room across from Grandma's. His bed had a high metal headboard. We'd climb up the cross bars on the headboard and jump onto the mattress below. It was great fun. I'll bet Grandpa was not so happy when he came to bed. It was generally really messed up.

One late May or early June, Jane, Jean, and I were together at Grandma's. The weather was warm and muggy. We had played enough Old Maid and Chutes and Ladders with Jean. Jane knew the small stream north of Grandpa's barn was flowing over some stones. We talked Grandma into letting us go if we took little Jean with us.

As we walked to the creek, the grass was knee-high to us and almost waist-high to Jean. We reached the creek and followed it to the fence flood gate line between Grandpa's land and the neighbor's land. There was a huge pile of concrete and rocks in the middle of the creek. The water flowed and gurgled over the stones. We tossed little sticks into the stream and watched them slide over the miniature stone waterfalls. We were oblivious of the approaching weather. The sky began to darken and there was the sound of thunder.

"We'd better get back to Grandma's house," Jane said.

We were about three hundred yards from the buildings. I started running until Jane yelled, "Help me with Jean. She can't run through this tall grass."

I turned and grabbed Jean's hand. Jane took the other and we bounced and dragged little Jean between us. The storm was catching us. I could see the lightning. The shelf cloud or wind cloud which precedes many thunderstorms was blowing the trees just behind us. We knew we couldn't stop. We about pulled off poor Jean's arms.

We were halfway when the first big drops started, but we couldn't go any faster. We were doomed to getting soaked. Suddenly, Grandpa appeared around the corner of the chicken house. He came running fast. His long legs carried him swiftly. He scooped up little Jean and ordered, "You kids get to the house."

I ran ahead and stopped under the eaves of the corncrib.

"No, get to the house, now!" he scolded.

We raced around the corner of the chicken house, across the gravel yard, and through the metal gate to find a worried Grandma waiting at the back door.

"Hurry! Hurry! Get inside. The rain is right behind you."

Just as we scurried inside, a big bolt of lightning cracked across the sky, followed by a loud clap of thunder. Grandpa hurried around inside shutting the windows. Grandma helped us out of our wet shoes and light jackets and led us to the kitchen table where she fed us cookies and milk. Jane and I were both worried about what Grandpa would say. He was a man of few words when it came to children

Grandpa appeared with a stern look on his face and said, "Next time check the weather a little sooner. You and Jane should have known the storm was coming."

Jane and I answered together, "We will, Grandpa."

Grandpa sat down and had cookies with us. He was relieved to have everyone safe. I even detected a slight smile as we devoured our treats.

One Sunday, Jane and I were alone again. As soon as dinner was over, the men retired to the parlor and the ladies to the kitchen table. I suppose Jean had to nap. The big three were doing something with the youth group at church. It was early March, so the outside weather was not pleasant. Jane and I headed for the basement.

Grandpa Shepard's basement was full of Pioneer seed corn bags as usual, stacked in rows of the different numbers identifying the variety and size of kernel. There were streets and alleys and little coves to play in. We climbed over the stacks and slid from pile to pile. We stayed in the room where the furnace was because it had better lights and seemed warmer. In one corner of this room was a half bath which contained a toilet and a lavatory.

Jane and I decided to play house. We pretended to wash dishes and other household chores. I ventured to the make-believe garden to pick a vegetable. Jane knew where Grandma kept some old blankets. We unfolded them on the floor of the little half bath. This would be our bedroom. We continued through our imaginary lives. It was time for bed. Jane suggested we sleep together like our parents.

She said, "We can't go to bed in our clothes. Mom and Dad wear pajamas."

"Yeah, I know, but we have no pajamas."

"Let's pretend we do. Why don't we take off everything but our underwear."

I replied, "Okay, you go first."

Jane didn't hesitate. She removed her shoes and socks. I followed. She took off her blouse and skirt. I followed with my trousers and shirt. We laid our clothes neatly on the sacks just outside the bathroom door. We kept on our undershirts along with underpants.

Now, mind you, our parents were just upstairs. We dashed into the little bathroom and slipped between the covers. Jane pulled the door shut. Just as she did, we heard the basement door open. Someone was coming down. Our clothes were lying outside on the sacks. Even if we had them with us, we didn't have time to dress. We laid there petrified. Who was coming down? Grandpa? Uncle Jim? No, it was Uncle Vernon.

He walked into the room and asked, "What are you two doing?"

"Just playing," we answered.

"Well, it is getting a little cool upstairs, so I came down to put some more coal in the furnace. Don't let me bother you."

"Oh, we aren't cold," Jane said as she pulled the covers up to her neck.

What if Uncle Vernon saw our clothes? How were we going to explain? We trembled under the blankets. Uncle Vernon stoked the fire and didn't say anything.

"You children have your fun. I think the fire will warm up the women. It's almost four, so you'd better wrap up your play. Your dads have chores to do."

"Okay, we'll hurry. See ya!"

Uncle Vernon turned and was gone. We waited until we heard the basement door shut at the top of the stairs and scrambled out of our make-believe bedroom and into our clothes.

We knew we were fortunate to escape this time, but we would never do it again. Our parents were too darn close. It was the last time I remember we ever got undressed while playing. We were quickly growing up and this childish playtime would soon end.

As Jane and I grew older, we played board games on Sundays. Many times, we included Jean. She was older and knew how to play some of the simpler games. Then, when Jane graduated from eighth grade—or was it after her freshman year in high school—I realized she was acting differently.

Previously, almost every Sunday afternoon in the summer, we went to the Muscatine swimming pool in Weed Park. It was a ritual. Most of my friends from church were there. We'd go there as soon as we could beg someone to drive us to town. We'd stay until three or three-thirty when our ride home would come. If we were lucky, we would have an ice cream cone from the park food stand.

At the start of a new summer and the first warm Sunday, Jane, Jean, and I made the trek to the pool. I jumped into the pool figuring I'd be followed by Jane. Instead, she wandered on the walkway around the pool with her towel in her hand. She was not interested in swimming and playing in the water. She sat with some girls along the fence to sunbathe.

I went over and asked, "Aren't you coming in?"

She looked up at me with my flattop haircut, and I must admit, a terrible-looking pair of swim trunks. "Not today, Bob. I'm going to sit here and visit with my friends."

The other girls just looked at me and giggled. I shook my head in disbelief. How could anyone not want to swim? What fun is there in just talking? I jumped back into the water and found some guys from church. The next time I looked at Jane there were some boys on the outside of the fence talking to her and her friends.

I didn't grow up that day, but I realized my best friend, Jane, had.

Uncle Vernon always was ready to try new technology. At one Sunday dinner, Sue or Lucy announced they had a television. Television was not unknown to us, but country folks usually had poor reception. We were at Grandma's when the girls told us all about the wonderful TV. There were all these shows on channels 4 and 5.

In mid-afternoon, Lucy drove all the cousins to her home. We sat down in front of a 17" television set. Lucy turned on the black and white screen. The first show was "Super Circus," followed by "Hopalong Cassidy." Mom and Dad arrived to take us home while we were still deep into the western. Dad watched a bit of the show and said, "We'll go home and milk. You kids stay here. We'll be back."

Later Mom and Dad returned to watch "Ed Sullivan" and "Colgate Comedy Hour." A few weeks later, Dad did his usual Dad thing. I'm sure Mom would have dragged her feet, but not Dad. McKee Feed & Seed sold televisions, so on a trip to purchase his supply of alfalfa, clover, and brome grass seed, Dad had the store pack a TV in the back of the truck.

Mom and Mary were at a Montpelier Merry Makers 4-H club meeting, so Dad and I assembled the big antennae, strung the lead-in wire through the window, and plugged in the set. It was a Capehart with five dials, on-off, volume, vertical hold, horizontal hold, and channel selection with twelve channels 2 through 13. Only channels 4 and 5 worked. As I saw Mary and Mom drive up the lane, I was excited to surprise them. Mary claimed I was literally dancing in front of the window.

The surprise wasn't much. Our garage was behind the house, so they saw the antennae planted in the back yard. But it was the beginning of a new era. Television became our new entertainment in the evenings.

Christmas with the Shepard family was like that of many other close-knit families. In the early years we always went to Grandma Shepard's. I was never excited about going, or at least, about the timing. We had to go right after Dad finished chores, since Mom had to be there early to help in the kitchen. I had just opened my Christmas gifts, and now I had to leave these new toys. Yeah, I knew Mom was Grandma's daughter, but couldn't Grandma do this by herself?

Once there I soon forgot my complaints. Jane and Jean would be there already. We were in charge of receiving the packages and putting them under the tree. By dinnertime, the whole family would be there. The Griffins would come from Letts. The kitchen was full of activity. I knew we were getting close to eating when Uncle Vernon started carving the turkey.

Dinner began with Grandpa giving grace. The cousins filled their plates and ate their meal on the sun porch. It was a sumptuous meal: turkey, mashed potatoes, green bean casserole, cranberry sauce, Grandma's escalloped oysters, plus other sides. The desserts were the best. Aunt Harriet's fizz, Mom's chocolate cake, mincemeat pie, and coconut cream pie, topped off by a plate of Aunt Florence's candies filled with those melt-in-your-mouth white divinity morsels. Of course, the cousins finished eating way before the adults. The big three had to help wash and dry, so we younger ones circled the tree checking for our gifts.

The system was, all the children received a gift from each family and grandparents. The adults drew names. You became an adult when you graduated from eighth grade. The gifts were distributed by Uncle Jim. I remember going around the room and watching each person open their gifts.

After all the gifts were opened, Grandpa would go into the adjoining bedroom and bring out another box or two of presents. These were from my Great Aunt Belle who lived in Spokane, Washington. She shipped her packages by Railway Express, and Grandpa went to either West Liberty or Wilton railway station to pick them up. Most of the gifts were unusual. Bill and I received play six shooters which had a mechanism to feed a tape of caps. The caps fed under the pistol's hammer and exploded when fired. We got to try one shot before they were banned until we got home.

Bill and I always opened our similar gifts together. Sometimes Jane would join in. One Christmas, we were on the last round of opening. Bill tore into his package and found a red steel dump truck. Boy, was it neat!

I looked at my pile and there was no identical package. I looked at my mother and was about to point out the discrepancy when she put her finger to her lips and indicated not to say anything. I kept quiet until we were headed home.

I asked, "How come Bill got a truck and I didn't?"

Mom answered, "Well, whoever forgot the package

will find it later and give it to you. I'm sure it was just mistake."

I accepted her response but still was not happy. I never got that red steel dump truck until much, much later — like, sixty years later. My cousins held a reunion in Vermont. We were there for a week. Every day was filled with stories and tales of our early days. One evening we got around to talking about the Shepard Christmas's. I related my tale of woe. Of course, I received much kidding after that.

Then Sue said to Lucy, "You know, I remember that red truck. It just appeared one day. We played with it. Mom must have forgot to bring it that day."

Lucy replied, "Yes, I remember it too. I thought at the time it was strange. We were twelve or thirteen and toy trucks were not something we desired."

I was kidded and teased about my red truck for a couple of more reunions until, at my farm retirement party, Cousin Jane presented me with a red steel dump truck. I finally got even with Bill. It proudly sits along with the rest of my antique toys.

When the little three reached high school age, the gift exchange ceased. The big three were in college, maybe married. It was difficult to draw names, let alone buy something for the receiver. Jean still complains that she and Anne were gypped out of a few years of the round-robin gifting.

Soon after Grandma passed, the Shepard Christmas dinners moved to other locations and more convenient days. We all had our families, our girlfriend's families, and spousal family dinners to attend. Those memories of the old time Christmases at Grandma Shepard's are still precious. It was when families were still important.

The last family Christmas when everyone attended except for Bill, who was in Florida, was held at my home after I was married. At this time both grandparents had passed away. We lived in an old farmhouse and the kitchen area was large. We borrowed tables from our church and had everyone seated at one long table. At the end of our meal, I showed a film my dad had taken of the Shepard get togethers several years ago. We all had a good laugh at our antics when we were young.

July 4th became a special holiday for a few years. When Bill moved to McAllen, Texas, he returned to Iowa in the summer and lived with Grandma Shepard. It was a fun time. Bill and Jane both could visit and stay over. Aunt Helen and Anne would fly up for the holiday. It was a long trip in those days because the fast jet planes were not used as passenger airplanes. All airplanes were propellor driven.

The celebration was always held at our farm. It was illegal to shoot private fireworks in Iowa, but the law was mostly ignored, especially out in rural areas. Heck, all you had to do was receive permission from the township trustee, and Dad was a trustee.

Our Fourths became much better because of Uncle Arlo. Because of his business, he could not take much time off, but he'd arrive for the Fourth and spend a couple of days with the family. Texas allowed fireworks, so Uncle Arlo arrived with a trunk full of contraband.

One year, we all attended the big parade in Muscatine. It was a typical July day, hot and humid. Instead of going to Weed Park after the parade for a picnic, the Shepards came to our home for a late lunch. Dad and Uncle Jim had milking to do, so there was a pause in the festivities. The party would resume around eight.

Jane and Bill stayed. Bill had a few incher firecrackers that he had packed in his suitcase. We put them under a frozen lemonade can and watched to see how high we could make the can fly. We tried tying two together for a bigger bang. The can flew twenty feet or more, straight up. Three tied together seldom worked. One cracker would always be late and be blown away. It was a waste of fire power.

The families returned from chores. We set up chairs on the front lawn and waited for Uncle Arlo. He had called Aunt Helen and said he was on his way. The sun set and evening arrived, but no Uncle Arlo. Aunt Helen was her nervous self. She worried her husband would not make it. There was no way of knowing where he was. Remember this is before cellphones.

Then a cloud of dust appeared down by the school. Could it be Uncle Arlo? The dust came around the curve west of the lane. We could see headlights. The car slowed and turned into our drive. Uncle Arlo's big Buick purred up the lane. He stopped out front and hopped out with a big grin on his face. He was the star attraction. He opened the car trunk, and the inside was loaded with skyrockets, roman candles, fountains, and hundreds of small firecrackers. In minutes, he and Dad put on the show.

Mother nature had a show of her own starting after the last skyrocket burst. To the west, the sky was full of lightning, but it was a long way off. We had television by that time, but weather forecasts were primitive and who would be watching TV now? We had ice cream and cake before going home. The storm was getting closer, and

Grandpa Shepard thought they should be getting home, because of the hot day, they had left the windows open in their house. Our party was over.

About an hour later, the phone rang. Mom and Mary were still washing some dishes. There were some that didn't fit in the top loading dishwasher. It was Aunt Helen. They did not get home in time to shut the windows. The thunderstorm they thought was further away beat them home. It was accompanied by high winds. Grandma's house was soaked from the driving rain. One bedroom was so wet they had to sleep on the couch. She explained across the road, many of Uncle Jim's lovely maple trees had toppled.

Mom asked, "Do you need us to come down and help?"

"No, we will be all right tonight, but tomorrow will be a major cleanup," she answered.

"We'll be down as soon as chores are finished."

The next morning, everyone, including our hired hand, went to help. Uncle Jim's neighbors were already there. His place had taken a major hit. Luckily, none of the trees fell on the house. Mom and Mary helped Aunt Helen mop up across the road. All bedding had to be washed and the mattress on one bed had to be carried out to dry. It was one of those wild Iowa storms which affect just a small area. Uncle Jim's corn field was flattened along with those of several neighbors to the north and east. It would be a Fourth we would never forget.

I do have cousins on my dad's side of the family. Dad's sister, Georgie, had two children, but they were several years older. In fact, their grandchildren were the same age as me. We had very little in common. Uncle Heinie and Aunt Georgie lived just one farm over. We would butcher hogs and cut wood together.

Most of Dad's cousins lived farther away. The Drumm family lived in McCausland, Iowa, and the Schiele family lived north of Stockton. They were Dad's first and second cousins. We would get together twice a year, once in the summer for a big Schiele reunion and at Thanksgiving where the Drumms and Bangerts attended. We didn't have the attachment of going to church and eating together each Sunday as we did with the Shepards.

My problem with these cousins was, again, they were mostly girls, at least when we met for Thanksgiving. The Schiele Reunion had some boys. At Thanksgiving Mary could team up with LuAnn Drumm, as they were close in age. LuAnn had three younger sisters, Jolayne, Marlyss, and Barbara. To me they were just three little girls. LuAnn's brother, Don, was close to ten years older than me. He and Wayne Bangert were pals. I was the outcast.

One Thanksgiving at Don's home, I was totally bored. He was a teenager and completely ignored me. Finally, his mother suggested Don show me his model train. Don rolled his eyes. He didn't want this little boy near his pride and joy. She insisted. He reluctantly consented and led me to his room upstairs.

His layout was primitive, but still fascinating to a young lad. It was an American Flyer train. His train was a S scale and ran on just two rails. He had it running under his bed and around the edges of his room.

Everything about his setup was what I wished for. He ran the train around the track, making sure I was not getting too close. Finally, to his relief, his mom hollered up that dinner was ready. We were hungry. Thanksgiving dinner had been delayed, and it was almost two o'clock. The table was filled with many delicacies. We always had roast duck, Aunt Georgie's specialty, and candied yams along with the usual entrees of turkey, dressing, and gobs of mashed potatoes.

This year we had our Thanksgiving dinner at the Drumms in McCausland. McCausland is a small town in northeast Scott County. The roads to it were not paved in the 1940s. While we were devouring our duck and pie, it started to snow. The wind began to blow and drift the white crystals. It was almost dark when we left. The Bangerts left first, and we followed. Before we reached the only paved highway in the county, the Bangerts' car slipped off the road and into a snowbank.

By the time we reached them, Wayne, who was old enough to drive by now, was searching the car trunk for his set of tire chains. Dad focused the car lights on the mired vehicle. Wayne struggled with the chains, but eventually wrapped them around each rear wheel. Once the chains were in place, Wayne easily drove out of the snowdrift. We followed the Bangerts to the paved highway where the chains had to be removed in order to travel safely. I don't know when we got home that evening, but I know Dad still had milking to do. He'd given the hired hand the day off.

Another time we ate at Great Aunt Annie's home. She lived in McCausland with Great Uncle Gus. The unusual thing about her home was that it had no running water. They had no indoor plumbing, only an outside privy. The elderly couple were accustomed to the way they lived and had no desire to change.

Mary and I made it through the entire day without going to the outside privy. Everything was cooked or warmed on an iron wood fired cook stove. The women pumped water from the cistern for washing dishes. Water for drinking was carried from the well outside. By the time we departed, Mom was exhausted, along with the others.

I remember going there only once for a Thanksgiving meal. I believe Uncle Gus died the next year and Aunt Annie shortly thereafter. I have journeyed to McCausland several times and found their home is still standing and lived in. I'm sure it has been fully modernized.

We had our Thanksgiving meals at different homes. It was at our home right after the house was finished. Dad set up tables in the basement. We had a gas range in the basement that Mom used for her canning, so some of the cooking could be finished downstairs. We didn't need to carry food from the upstairs kitchen. Dad lit the fireplace in the basement and showed off his new domicile. He was so proud of his home. It was his dream.

As with the Shepard family, the children grew up. Going to the different family events was not required. The older generation passed on and the traditions faded. Soon our time with Dad's cousins was just a memory, though I still see Jolayne, Marliss, and Barbara Drumm occasionally. Of course, they are all married and have different last names. I have found I have become much more interested in the Bancks side of my family as I age. The Drumm girls, as I call them, are part of my Bancks family heritage.

*

We had another group of friends who were not related to us or to one another. They were just four couples with like interests. Ed and Helen Plett lived a mile away when my parents were first married. They and my parents played the card game 500 once a month during the 30s. At some time the group expanded to include the Baker family. I have no idea of when or why or how this happened. They must have decided to play card games, but they needed a fourth couple. Lester and Freda Chambliss moved in a half mile from the Bakers. They were somewhat younger but that didn't seem to affect the group.

The four couples met every summer and every New Year's Eve. The Bakers lived on the family farm as we did. The Pletts and Chamblisses were renters and moved several times before being able to purchase a farm. The New Year's Eve parties were the best. Each family would take a turn hosting.

When the Chamblisses lived near the Bakers, the road to their place was not graveled. This party night was New Year's Eve. Lester still owned a team of draft horses. For a treat he hitched his team to his bobsled. Inside he lined the sides with bales of straw. Everyone climbed in and wrapped themselves with blankets. It was very cold, probably below zero.

We started on our ride down the frozen mud road. When we approached a neighbor's house, we would sing "Jingle Bells." For some reason, Dad was designated to ride on the last bale. Somehow, he jumped out or slipped out of the sled. He began to running to catch up. Lester thought he was just trying to warm up, so he didn't slow down or stop. Everyone but Mom thought it was funny.

Finally, someone convinced Lester that Dad was not bluffing, and he needed to stop. Dad climbed aboard and rode the rest of the way. He received a lot of teasing when we got inside the Chambliss' home.

The New Year's Eve parties started with a big meal. The men would play cards while the women visited. There were seven children, and board games were our entertainment with ages ranging from sixteen to two. At the stroke of midnight, we'd all yell, "Happy New Year." Some of the kids had noise makers but the noise was suppressed quickly. Before we went home, there was another midnight snack, which was more of a feast. Many times, someone had cranked out some homemade ice cream. We'd start home about two in the morning. I guess Dad still milked the cows on time the next day on three hours sleep.

Don Plett, the younger of the Plett sons, was seven years older than me. I loved to go to his home. He had a collection of farm toys which boggled my mind. If some of the equipment was not available to buy, he was talented enough to make his own from parts of his old ones. I went to his place just to dream. At one of their farms, he had an entire room for his collection. His mother knew I wanted to play with some of his toy tractors.

She'd say, "Donald, you get Bobby some of your tractors to play with. He'll be careful." Poor Don would pick out a couple of pieces which weren't his favorites and let me have them. All the time I played, he watched with a worried eye. I never broke a piece.

At home, I would try to emulate his collection, but hard as I tried, I never accomplished his skills.

The other great place for our parties was at Lester and Freda's last farm. Freda's parents retired and sold their farm to Freda and Lester. The house was fantastic. It was state-of-the-art when built since it had indoor plumbing. And it was huge, with two staircases plus a bathroom upstairs. The toilet in the bathroom had a tank mounted high near the ceiling. When you needed to flush, you pulled a wooden handle hanging on a chain.

The front staircase held a stained-glass window at the first landing. There was a living room, a parlor, a dining room, and kitchen downstairs. Upstairs there were four bedrooms, plus a room for the hired girl. Outside, the house was encircled by a porch. Freda's parents were apparently wealthy.

The group stayed together until Dad and Mrs. Plett died. Then Mom couldn't go to the gatherings anymore; there were too many memories. I guess Dad and Helen were the glue which held the group together. Of the seven children of this card group, I'm the only one still around.

Travels with Dad

MY DAD WAS an adventurer. He loved to travel and see the world. He would conjure up ideas and events to attend. I got to see many places in our country because of his passion. Our family went on several trips, starting with his pushing my uncles to do the extraordinary.

He'd say at a Sunday dinner, "Say, the Tulip Festival is next weekend at Pella, Iowa. Let's go."

The uncles knew there was no way to turn their brother-in-law down. He was persistent.

After some quick conversation, they decided to go. The Tulip Festival was in May, prime time for field work, so Saturday was out. The Shepards never worked on Sunday except for chores and maybe loading out cattle for Monday's market in Chicago. Sunday was the only day available.

The next problem was, Mom and Grandma said we had to go to church first. They were good Methodists. I believe we left church right after Sunday School, which was first. Jane, Jean, and I had to keep our attendance perfect. We skipped out on the regular service.

Pella was over one hundred miles away. Interstates didn't exist. The highway to Pella was US 6, which came within a few miles of the Dutch community. Dad had read about a former New York Yankee baseball player, Bill Zuber, who opened a restaurant in Homestead, one of the villages of the Amana Colonies.

The Amana Colonies was founded by a group of people from Germany. They established an almost communal society. They had a huge farming area, woolen mills, and many craftsmen who built furniture. They were famous for the homecooked family style meals. We arrived in time for lunch, but because our group numbered fourteen, we waited quite a while to be served.

We made Pella later in the afternoon. It wasn't quite the festival it is today, but the tulips were beautiful. Jane and I ran and skipped around but never tiptoed through the tulips. I have no idea when we returned home, but chores still had to be finished. Mom fixed a late supper.

This was just one of Dad's many excursions where he included the Shepards. Once he found out they too enjoyed the adventure, he planned more. I think he had fun showing others the world beyond Muscatine County. I received the benefits of his adventures.

Dad dreamed of going to the State Fair for a couple of days with the Shepards. This would not be unusual except he wanted to camp in the campground. Everyone would be together for meals and sleeping. Motels were small and few. Most fair goers stayed in the hotels in downtown Des Moines. The only members of the family who didn't go were Grandpa and Grandma and baby Jean. Even the Griffins attended. Anne either wasn't born yet or she stayed with Bill's other grandparents.

It was quite an undertaking. There were fourteen of us. We started with Uncle Jim's truck. The men threw a huge tarp over the livestock rack. Inside they placed old mattresses. Uncle Jim had one of his hired hands construct steps up into the truck bed. This would be the women's quarters.

The men erected Army cots and slept outside. In case of rain, they had another canvas stretched out from one side of the truck. Jane, Bill, and I were to sleep in the seat of our Studebaker. One on the front seat, one on rear seat, and one on the floor in the back.

We embarked late morning with cases of soda, coolers of water, and ice chests of all kinds of food. The caravan started for Des Moines. We set up under a big oak tree in the fabled State Fair Campgrounds, assembled tables and chairs, and after a quick lunch, it was time to see the fair.

The state fair at that time leaned toward agriculture. The big tractor companies were ready to display the new improvements in equipment. There were tents set up for the farmers and demonstrations were presented. I believe this was the year John Deere introduced its Roll-a-Matic front wheels.

I haven't a clue how the women entertained themselves. The big three took Jane and ventured off by themselves. Bill and I followed the men. After all, the tractor companies sold toy tractors and implements inside their tents. Maybe Dad would buy me a new toy tractor.

A miniature steam train ran through the grove in the center of the fairgrounds. The little three got our thrills there. We did ride some of the quieter rides on the Midway as well.

The ride we could go on with the big three cousins was The Old Mill. We got into a little wooden boat and floated through a maze which was covered with a tight black roof. Once you were a few feet inside, it was pitch dark. Once in a while, an eerie scene appeared. I sat with Lucy. She put her arm around me as I shuddered in the pitch-black tunnel. But when we emerged at the end of the tunnel, as any kid would say, "Let's do it again." In those days it only cost a dime to ride.

The next day we attended the car races. Although we were comfortable and I'm sure the adults enjoyed the race, it was boring to me. In the evening we attended the grandstand show. It was not a big-name singer or band, but a Las Vegas-type show with singers, comedians, and pretty girls. The girls were dressed in fancy costumes, which wasn't unusual, but they came out of a large pool of water on the stage. They emerged dry as a bone, did a dance, then disappeared back into pool. Some other performers did a number followed by the dancers emerging again in completely different costumes.

Each time they emerged they were wearing a little less, and I remember hearing Dad saying, "I wonder how many times the girls can disappear and re-appear and still be wearing something?"

Good thing my mom didn't hear him.

Aunt Florence had this phobia that we all had to be regular. She'd go to everyone and request we drink a small cup of prune juice. No one escaped her except Bill. It seems when Aunt Florence came close, Bill was nowhere to be found. I think he and Uncle Arlo had a plan to avoid Aunt Flo.

On the third and last morning, Dad suggested the whole family breakfast at Hardenbrooks Restaurant located at one end of the Cattle Barn. They were known for their endless servings of light and fluffy pancakes.

This went over well with everyone except Aunt Florence. She claimed pancakes would settle and we'd all become constipated. We went anyway. We sat at a big picnic table and the waitresses brought pancakes and bacon. They were delicious. I don't believe anyone became plugged, so, sorry, Aunt Florence.

We broke camp later that morning and headed home. We were tired and probably dirty, but what a great time! We wondered what my dad would dream up next.

As the cousins aged, Dad had another brainstorm. He said, "Let's go to Chicago for Thanksgiving break. The world-famous International Livestock Exposition will be in session. Michigan Avenue is decorated for Christmas. There is plenty to do in Chicago."

The Shepards agreed and plans were made. Dad used the livestock show as the carrot to get the Shepard men interested. He had bigger ideas. Mom's friend from high school, Alpha Braunwarth, was familiar with Chicago. She and Dad planned the rest of the trip for the whole group.

On Thanksgiving, we drove to Chicago and checked into the Morrison Hotel downtown. That evening we walked the Miracle Mile on Michigan Avenue. All the Christmas decorations were displayed, and the windows of Marshall Field's depicted a fairy tale with animated figures moving with a series of gears and chains.

The next couple of days were a whirlwind of activity. First it was off to the Museum of Science and Industry, followed by a session at the Adler Planetarium. It was getting late, and Jane and I were to visit Santa in Marshall Fields, and when we got there, Santa was on break. We'd have to settle for seeing Santa in Davenport or Muscatine. Later that evening, Dad had another surprise. We went to Shubert Theater and saw the Broadway musical *Annie Get Your Gun*.

On Saturday, before we journeyed to The International Amphitheater, we had breakfast at an unusual café named The Ham and Egger. It was a breakfast venue recommended by Alpha. Inside the walls were painted with murals of pigs, chickens and eggs flying all over.

The Amphitheater was located near the Chicago Stockyards in South Chicago. Getting there by car would be a problem, and so would parking for the day. Dad thought we should ride the street cars to the amphitheater. He got directions from the hotel desk and off we went.

The streetcar popped and rattled as we rode. There were many stops as people got off or on. As we got closer to the stockyards, the car become more filled with dark skinned people. We were on the south side of Chicago. We arrived just outside the Amphitheater. The place was huge. There were hundreds of cattle and their owners from all over the U.S. and Canada, all under one roof.

We walked around the cattle holding area and discovered Walter Baker and his family tending their cattle. Walter was Aunt Harriet's brother, and his family made the annual excursion to the show to show their animals. In the afternoon, we watched an equestrian competition where they jumped horses over gates and rails. It was very boring to an eight-year-old boy.

The ride back to the hotel was long. By this time, I was exhausted. The streetcar was crowded. I was about to sit on Mom's lap when a large dark-skinned lady across from us offered to let me sit next to her. There was just enough room for a kid. As we traveled, I became very sleepy. Mom touched me a couple of times, but soon I just couldn't keep awake. I lay my head on the lady and fell asleep. Mom said she was about to wake me, but the lady shook her head no. She put her arm around me and held me until our stop arrived.

Mom said, "Thank you. I'm sorry, I didn't mean to have Bobby bother you."

The lady replied, "Don't worry, honey, I have five children at home. It didn't bother me a bit. Have a nice evening."

Uncle Jim thought it was funny how I slept on the lady's lap. He kidded me for years about the incident.

In the evening we met Alpha at Don Ho's Restaurant. It was a night club with a South Seas theme. Fake palm trees lined the walls. The food was all South Pacific cuisine, and Jane and I were not too enthused with the menu. The only food we ate was steamed peas in the pod and pineapple chunks. I guess there was no kid's menu at a night club.

The next morning was Sunday. Of course, we had to go to church. Jane and I didn't want to mess up our attendance record back at Sweetland. There was a large Methodist Church named the Chicago Temple within walking distance. It had a vast auditorium. Jane and I made sure we secured our church bulletins to prove we had attended.

After church, it was homeward bound. I'm sure we had lunch somewhere along the way, perhaps in the town of Walnut. The uncles stopped there often when they delivered cattle to the stockyards.

My going to Chicago opened my eyes to a place where many, many people lived. Everybody seemed to be in a hurry. Riding the streetcar showed me how people lived all crowded together. There were no open spaces. I felt sorry for the children who had to live there.

Dad's escapades abated for a year or two when he became involved in building our house. By the time it was finished, Uncle Arlo's family had moved to McAllen, Texas, which is located at the very southern tip of the state. All the Griffins did was give Dad an excuse to go traveling again.

Schools never started until after Labor Day, so there was a two-week gap between the West Liberty Fair and school. Dad proposed the adventure to the Shepards again. Uncle Jim said he couldn't be gone that long from the farm. Uncle Vernon said yes since he had given up dairying. We planned to leave right after the fair, until my sister threw a monkey wrench into the plans.

In girl's 4-H every year, there was a competition involving demonstration teams from each 4-H club. Two girls would demonstrate the proper way to bake, sew, or decorate. This year Mary and our neighbor girl, Marilyn Watts, represented the Montpelier Merry Makers 4-H club. Their presentation was on baking a yellow butter cake. The practice paid off, for the team won the county competition. They would represent Muscatine County at the State Fair.

This screwed up Dad's plans. If we stayed to see Mary at State Fair, there would not be time to journey to Texas. A deal was made. Mary would not go to Texas. She would stay at Aunt Harriet's and compete at the state fair. Jane would replace Mary on the trip south.

I loved the idea. It would be a whole two weeks with Jane. If Mary went, she, Sue and Lucy would be pals and I would be a loner. Mom was not happy with this plan, but she decided to follow along. She wanted to see her sister Helen. Dad was not to be denied his trip.

Texas is a very big state and there were no Interstate highways. Our days were long and hot. We had no air conditioning in our cars. We took a photo of the sign saying, "You are now entering Texas." It was in Texarkana.

Our first major stop on the way to McAllen was Houston. The Shepards had an acquaintance who lived there. Dad wanted to check out the famous Shamrock Hotel in downtown Huston. He wasn't going to stay there, for this was a five-star hotel, built by an oil wildcatter who had struck it rich. We just walked through the lobby and stared at the gold and glitz of the building, then stayed nearby.

The family we were visiting invited us to come in the afternoon. We had the morning to kill. I don't know what the women did, but we men went to the hotel barber shop for a shave and a haircut. I didn't need a shave, but I did need a haircut. The shop was a room with several chairs and mostly African American barbers. When my turn came, a man with gray hair set me on a board to raise me to his level. He asked the usual questions about where I lived, and so on. I felt a little nick.

"Oops!" the barber said. "I nicked your ear. I'm sorry. Does it hurt?"

I shook my head.

He kept apologizing to me and Dad. It couldn't have been very bad because he stopped the bleeding with a piece of tissue paper.

Dad paid him and said, "Don't worry. He'll be fine. He's a farm boy and one little nick won't hurt him."

I believe the barber was afraid my dad would be angry. Maybe some white people would have been upset, but I was used to scratches and cuts. To Dad and me, this little incident was not going to affect our vacation.

Our afternoon visit was to a friend's home. They had two teenage sons. I suppose they were instructed to entertain me while the adults visited.

One of the boys asked me, "Would like to see our model train layout?"

I answered, "Yeah."

They led me to their bedroom where they had assembled the most fantastic HO model layout. It had little trains and buildings and mountains. Their room was filled with the set-up. I was amazed. I gingerly walked all around the layout, being careful not to touch anything. It was fabulous. This layout was built before all the modern-day kits and track were available. In time the rest of the family crowded into the room. They were impressed, but not as much as me. I could have stayed there for hours.

We journeyed on to McAllen. Aunt Helen was glad to see Mom. The Griffins had a spacious home with a large screened-in porch. This was the base of our operations for the next three days. We must have arrived on a Saturday because Bill, Jane, and I attended Bill's Sunday School class the next morning. Jane and I had to keep our attendance record clean.

Uncle Arlo took everyone else to visit a former hired hand of the Shepards, who now grew cotton in the Rio Grande Valley. In the afternoon, we traveled to the Gulf Coast and South Padre Island. There, you could drive your car right on the beach. There were no hotels or restaurants, just a wide expanse of low, flat, hard sand.

Jane did not bring a swimsuit, so before we headed out, Aunt Helen conjured up some shorts and a blouse for her to wear. We had a great time dodging the waves. Even the adults removed their shoes and socks and waded in the surf. It was late when we returned. Uncle Arlo wanted us to go to his favorite restaurant.

Because we had to shower to remove the sand and salt from the Gulf, it was late before we reached the restaurant. The venue was noted for all the boiled, steamed, and fried shrimp you could eat. By this time my palate had matured enough, and I could enjoy the cuisine. I gorged myself.

It had been a very long day, and we were the only patrons in the place. I wasn't too sharp at this hour. I had to go to the restroom, so I excused myself and walked off. When I returned, everyone was giggling at me.

"What's so funny?" I asked.

Bill pointed at the restroom sign. I had entered the women's restroom. Lucky there were no ladies in there at the time. It would be another one of my goofs I would be kidded about for a long time.

Uncle Arlo wanted us to experience Mexican culture. Dad, of course, was with him. The two men decided to go to Monterey, Mexico. It was a long trip. The city was very tropical. We walked around the marketplace and saw streets full of children begging for pesos.

Dad and I went shopping. I found a baseball bat which was carved and stained with Mexican figures. Dad bought a pair of loafer shoes for eight dollars. He figured he got a heck of a deal. He bragged about them for months.

We ate at Sanborn's Restaurant before returning to Texas. It was one which Alpha had recommended since she had been there before. Many restaurants in the city were not inspected for cleanliness. Jane and I had our first authentic tamales wrapped in real corn husks.

The next day we were supposed to be heading home. I don't know if Dad had already planned the extra miles before we visited McAllen or it was one of his spur of the moment decisions, but he figured he had enough time to see a major wonder of the U.S., Carlsbad Caverns in New Mexico. It was eight hundred miles away, but I think he thought he'd never be any closer. He liked spur of the moment ideas.

Dad said to Uncle Vernon, "I think I'm going to Carlsbad."

Uncle Vernon replied, "I think we'll return home."

Dad replied, "Okay, but I'm going to Carlsbad. It's one place I have never been, and I've read it is fantastic. See you in a week."

We headed west and Uncle Vernon headed north. The highways in Texas are wide and straight. There were no speed limits. Most Texans drove big, powerful cars and eighty miles per hour was the norm. It was close to eight hundred miles to Carlsbad Caverns, but miles fly by when you are traveling at eighty miles an hour. Dad's new Studebaker was performing great.

Jane and I napped in the back seat. The wind flowed through the car. Mom dozed. Dad slowly pressed the accelerator, 85, 90, 100 mph. The Studebaker screamed across the Texas landscape.

Mom woke up and hollered above the roaring wind, "Carl, what are you doing? How fast are we going?"

"We just hit a hundred about fifteen minutes ago." he replied with a grin.

"Well, slow down. Now!"

By this time Jane and I awoke. We were enjoying the speed. The hot wind had almost fried us in the back.

We were back down to eighty when Dad looked into the rearview mirror. He said to Mom, "Look who is coming up from behind."

Mom turned around and so did Jane and I. Gaining on us was Uncle Vernon's big green 98 Oldsmobile. The Shepards had changed their mind. I can't imagine the speed Uncle Vernon was going to be able to close the gap between us, but we made Carlsbad that evening. The next day we toured the famous cave. It took all day. We even ate lunch in the cave. Dad had made his goal. This was just one more sight checked off his mental list.

Crossing the Texas panhandle in August was hot. That night we stayed in Amarillo, Texas, in an older motel right off the famous U.S. Route 66. Motels were usually individual units. Our little unit was okay. The one where Uncle Vernon's family slept was not. The next morning Lucy told us of the huge cockroaches crawling up the walls. They did not sleep well.

Somewhere east of Amarillo, the highway department was resurfacing U.S. 66. There was no way around the delay. We waited in the heat for a pilot car to lead us through the construction zone. Jane and I were bored. We asked if we could just step outside for a minute. Mom thought it would be okay. We were barefoot. Both of us exited the car and jumped right back in. The asphalt was so hot we both burned the bottoms our feet.

We made it home a day or two later, stopping at Uncle Jim's to pick up Mary. She'd had a good experience at the Iowa State Fair. She stayed in the girls' 4-H dormitory and met other girls from all over the state. She and Marilyn had received a big blue ribbon. Dad was happy because he had felt guilty for not changing his plans to watch

Mary at the fair. They both had a good time. So, this ends the adventures with the Shepard families.

In time, the cousins grew up. Our parents insisted we all attend college somewhere. College spread us out further and contact between us was seldom. The only time we were together was maybe Christmas or a funeral of an aunt or uncle.

The cousins as a group fared well. Lucy became a nurse with several degrees. She finished her career at the University of Arizona as head of the nursing department. Sue received her doctorate in mathematics and headed the math department at Southeast Missouri State. Mary became an elementary teacher with a master's degree in children's literature. She taught school for fifty years. She must have been well liked because at her retirement party, fifteen hundred parents and former students attended.

Jane was also a teacher and obtained her master's degree in Special Education. Bill is an attorney in St. Petersburg, Florida. Jean became a teacher in middle and high school. She also has a master's degree. Anne worked with special students and had a master's degree, and I graduated from Iowa State with a Bachelor of Science degree in Agriculture.

Of the eight cousins, six of us are still alive and in our eighties. Although we were apart for many years, we now gather once a year to talk over the fun times we had. We still are family.

Our Family Adventures

DAD ALWAYS HAD a desire to see the world. He had to wait until I was almost six before our first vacation, since Mom thought I was too young to travel before then.

We loaded up the old Studebaker and headed West. His first destination was Colorado Springs. There were many attractions there. Beside Pike's Peak, he knew of a shrine dedicated to one of his favorite entertainers, Will Rogers. There is a tower called The Shrine to the Sun built in Rogers's honor. We climbed to the top following spiral stairs. On the walls were photos and news articles about Will Rogers. Once on top we could view the magnificent Rocky Mountains. I remember it being very crowded. People were going up and down the same staircase. It was slow. I hid behind my dad on the way up to avoid being crushed by the crowd. Coming down we were on the inside of the stairs and went much faster.

We followed a canyon to a tourist site called Seven Falls, a cataract of water tumbling down a cliff. I bought my first ever souvenir, a little trinket with a photo of the falls. The very next day Dad took the challenge of driving to the top of Pike's Peak.

The road was gravel and full of curves. Once we made it to the top of the 14,110-foot peak, we shivered in the cool air. None of us had dressed warmly enough for the altitude. We dashed inside the visitor center and had some warm cocoa. I got my second souvenir, a sailor hat with Pike's Peak written on the side. This was like a second Christmas. Before we started back down Mom snapped a photograph of Mary and me standing on the cog railroad platform.

Somewhere along the descent, the Studebaker decided to break down. Something happened to the motor. We were stranded. A car stopped, and the driver said as soon as he reached Colorado Springs, he would send a wrecker up to us. We could only hope he was successful.

We sat for a while until another tourist stopped. He had to be another farmer because he had a log chain stashed in his trunk. He attached his car to our front bumper and began pulling our car. Back then cars had real bumpers attached to the frame of the car. The Studebaker had mechanical brakes and didn't require the engine to be running to function, so Dad kept just enough pressure on the brakes to keep the two cars apart. We were almost to the bottom of the mountain when the wrecker arrived. Somehow, we got back to our motel, probably by taxi.

We spent the next two days in Colorado Springs while the Studebaker was repaired. Dad found an amusement park, though most of the rides were shut down, except the boat ride. Mary and I rode the little boat around in a canal. Because we were the only customers, we got to float around a couple of times. Later we toured the small city zoo, which today is much larger and known as The Cheyenne Mountain Zoo. Finally, the garage called, and we were once again on the road.

The next stop was Denver. We visited another one of Dad's finds, Buffalo Bill's Retreat. It was located on Lookout Mountain just outside Denver. I had heard of the famous man and his many feats but I was still too young to grasp the significance of his home. To appease me I got to purchase a little leather coin purse. Man, I was making out like a bandit.

Dad's final stop was Estes Park, Colorado. I'm sure we drove into Rocky Mountain National Park, though I don't remember anything but more mountains. I recall the drive down the Big Thompson River Canyon with its sheer cliffs towering over us. The river roared alongside.

Two days later, we were home. I didn't have time to digest our trip because the next day was my first day of school. Actually, school had already started and we were a day late, and Mary had only two others in her class. She wasn't far behind, and as I was the only student in my class, there was no one to catch up to.

We had survived our first trip. Dad proved to Mom I was old enough to travel. There would be many trips to come. I'm sure he planted in me the desire to travel and visit the many wonderful places in the world. Dad thought it was important to understand other people and places.

Dad's ideas for travel came mainly from two sources. The first was a travel magazine named *Holiday*. It described the many wonderful places in the world. I'm sure my father was planning on traveling abroad as soon as Mary and I became independent.

The second source of Dad's ideas was Mom's high school friend, Alpha. She was single and a professor of languages at Ball State University in Indiana. She and Dad planned many of the escapades for the Shepard family. Her travels took her around the world, and she carried a 16mm film camera with her.

The camera was great, but it didn't lend itself to her classroom presentations. Slides and still photographs were much better, so she upgraded to a new 35mm camera. One evening, Alpha came to visit. I was always excited to see Alpha because she brought Mary and me a book from Chicago or Muncie. This evening she brought me nothing.

When bedtime came, I trudged off, but not without blurting, "I didn't get anything from Alpha."

Mom was shocked and embarrassed. "Bobby, that is bad manners. Alpha does not have to bring a gift for you every time she comes. Now off to bed. We'll talk about this in the morning."

I knew by the tone of her voice; I was in trouble. The next morning Mom had calmed down. She led me to a cedar chest in their bedroom. Under it was a stack of phonograph records Alpha had left. These were 78s and mostly classical music. I guess Alpha didn't appreciate the modern stuff.

The most important gift was the 16mm film camera she gave to Dad. Mom told me Alpha worked late into the night teaching Dad how to operate the camera. From then on, many of our family adventures were recorded on film. Without the camera, my recall of our trips would be vague.

Next year's Yellowstone trip started with a stop in South Dakota to visit the Thoenes, another set of Dad's cousins. Dad's mother was a Thoene. The South Dakota cousins were close to my dad. My parents spent their honeymoon with Fred and Lillie Thoene. The four of them traveled to California in 1929 in a Studebaker. The roads were not developed, and tourist accommodations were few. It is almost unbelievable what they accomplished.

On our trip Dad wasn't sure every night we would have lodging, so he tied a rolled-up canvas on the rear bumper in case we got caught in the wilds. His plan was I could sleep on the rear seat floor while Mom and Mary slept on the seat. He would sleep outside on a cot. The canvas was just in case it was rainy. Thank goodness, he never had to use the system. Motels and hotels had improved quite a bit since 1929.

We traipsed through the Badlands of South Dakota and Wall Drug, then Mount Rushmore and Deadwood. We drove to the cemetery where Calamity Jane and Wild Bill Hickok are buried. Dad had read about the site in one of his travel magazines. We toured underground at Wind Cave and stopped to take some film of some wild burros who begged tourists for food.

After a stop at Devil's Tower, we arrived at Yellowstone. We stayed in a cabin inside Yellowstone National Park. The Park furnished bedding and firewood. Dad fired the little stove to keep us warm. He heard someone say the beaver were working a short walk away. We went to see, but the crowd noise scared the beavers, and they retreated to their lodge.

In the middle of the night, we were awakened by something tossing garbage cans around. Dad peered out the rear window and saw three black bears rummaging through our garbage. Good thing it was night, or he might have tried to film the bears.

While seeing the many wonders of the park, we came upon two bear cubs in the road. The line of cars stopped. Dad saw it as a photo opportunity. We found some stale breakfast rolls. Mary and I exited the safety of the car and hand fed the cute little cubs. Now, as I watch the film, I shudder to think how dangerous it was for us to do be outside the safety of the automobile. The mother bear could have been just over the rise. I guess we stupid farmers from Iowa were lucky.

We waited for Old Faithful to erupt so Dad could catch it on film. Dad went through a lot of film cartridges on the trip. Instead of heading home, we journeyed north to Glacier National Park and, on the way, the site of the Battle of the Bighorn where Custer was defeated. In 1948, there was only a monument surrounded by a wrought iron fence, not very impressive, but Dad had read about the famous battle, and he wanted to see it for himself. It was part of his research which drove him to explore.

At Glacier, Dad made a phone call home. When he returned from the phone booth and the long-distance phone call, he said, "Wayne said we had a hailstorm yesterday. Some of the hailstones were as large as eggs. He put some in the freezer to show us when we get home."

Mom asked, "Then we are heading for home tomorrow?"

"Why? Wayne said he put storm windows over the broken glass on the west side of the house. The rest of the house is okay. The corn is so beat up it will be a mess to pick. We can't do much for the crop now. He said he didn't see any reason to hurry home. I read about Banff and the Columbia Ice Fields near Calgary, Canada. It is only a day or two drive from here. Let's go."

I think Dad had planned this extension of the trip all along. Alpha had planted these sites in his head, and he knew it might be years before he got back to this area. So, we went to Canada. Passports were not required to enter, just a valid driver's license. We stayed in Calgary and shopped at a huge Hudson Bay Department store. Inside was the first escalator I had ever seen. While the rest of the family shopped, I rode the escalator up and down many times. Again, the environment was different than today. I was left alone to ride the escalator, and no one seemed concerned.

When we reached Banff late in the afternoon, it was cloudy and raining. Every motel had a No Vacancy sign. Mom was getting very nervous. Dad might have been nervous, too, but he never gave up. We pulled into a motel which was a group of little log cabins. The sign said No, but Dad went to the office anyway. We sat in the car wondering where we would sleep that night.

When Dad returned, he was all smiles.

"They found a unit for us for two nights. It is just up the hill."

Mom was relieved. We drove to the log cabin and entered. It was very nice. I mean very, very nice. Probably a cabin which was rented for weeks or months at a time. It had a complete kitchen, three bedrooms, and a living room with a fireplace. Soon we were relaxed and warm. I don't how he did it or how much he paid for the cabin, but we were safe and had good beds. I didn't have to sleep with my sister since we had our own rooms.

The next day we planned to visit the wonders of the Banff area. Dad had read about the Columbia ice fields and Alpha told us to visit Lake Louise.

The famous hotel with its flower gardens were built for the tourists coming from the east. It was something just to walk through the lobby or eat in the restaurant. Of course, Dad had to have a meal there despite Mom's objections.

She said it was too expensive. I don't think we ordered very much.

Dad followed Alpha's guidelines. In Kicking Horse Canyon, we watched a locomotive circle inside a mountain. We stopped at a lakeside hotel. Dad wanted to take a ride on the boat that toured across the lake and up the river outlet, but we were late and missed the last boat for the day. Dad decided to rent a small boat. He would take us all on a personal ride. Mom refused to go. She thought it was unsafe. Mary sided with her mother.

Dad said, "Okay, you go sit in the car while Bobby and I ride the boat."

We took off from the dock and waved at Mom and Mary. About an hour later, we found them sitting in the car. Mom was in tears. She apologized to Dad for refusing to go with him. In true Dad fashion, he told her to dry her tears. It was over. Tomorrow would be another day.

On the way home, one of the motels where we stayed was a group of old passenger train cars. The owner had converted them into living units. I made friends with the owner's son. We played with the little cars and trucks I had brought along. He liked my plastic toys and asked if I would trade with him.

A bargain was struck. He received two of my Banner trucks for his Canadian metal bus and car with fiber wheels. I don't know if all Canadian toys were this way, but fiber or cardboard tires don't work very well after they get wet. I got the short end of the deal, but I also felt sorry for the Canadian children because they didn't have access to toys with wheels that turned.

I don't know how far north we were, but Dad made a point of reading the newspaper outside at eleven o'clock in the evening. He also spotted a moose on the adjacent railroad tracks. He grabbed the camera and started shooting only to discover later his thumb was over the lens and he got no photos. He always called those shots "buck fever." That's a hunter's term for missing a shot.

Mom didn't believe in eating at restaurants all the time. Somehow, in our crowded car trunk, she found space for an ice chest. We would stop in towns and get supplies for noon meals. On our way back across Montana, we were far from any town. There were no roadside parks on the map, either. Dad knew we were getting hungry and began looking for a tree or abandoned farmstead to park and have lunch. He found a one-room school, much like Hazel Dell #3 back home.

We pulled in and found some shade behind the building. He tried the hand pump at the water well and it worked. We washed up and huddled out of the dry wind to eat. Like all one-room schools, outdoor privies were available. We used the facilities and started to pack up.

Dad, being the adventurer, decided to try the door to the building. It was unlocked. He went inside and found written on the blackboard, "Welcome to our school. You may use our building for a rest. Please leave it as you found it and leave your name on the blackboard."

The board was covered with many names of families and travelers from all over the country.

Dad signed our named and said with a smile, "Next time I'm checking the door before we eat."

We made it home a few days later. We needed to stop at the Thoene Ranch for Dad's South Dakota fix. It was threshing time and Dad just had to film the big threshing machine and Great Uncle Julius watching the men work.

Dad's insatiable desire to travel never waned. Next year he wanted to see Washington, D.C. He had another cousin, Royal Rostenbach, who lived in Arlington, Virginia, and worked for the government. Royal and his wife, Margret, visited Davenport every summer. His mother, my great aunt Ivy, lived there. He invited our family to visit them in D.C., and Dad decided to take him up on his invitation.

We stayed in a hotel near the Rostenbachs because they had no children and lived in an apartment. They showed us the many sights of the Capitol city. Since Royal had to return to work on Monday, Margret became our tour guide. She took us to the memorials, Arlington National Cemetery, and Mount Vernon.

It was at the National Gallery of Art where she showed her pluck. At that time, children under ten years of age were not allowed in the gallery. I was nine. She ignored the sign and walked right past the guard.

"Look like you are someone important and they will let us alone," she ordered.

We didn't have much time to spend, but she knew where the precious art was located. All the while we were there, the guard followed us. I guess he thought I might break something.

The next morning Royal took off from work. He insisted we eat breakfast at their apartment. When we got there, Royal was fixing breakfast, hand squeezing oranges for orange juice. He prepared a great meal. It struck me as odd because I had never saw a man fixing breakfast. Mom always made ours. Dad cooked a little when Mom was sick or away helping Grandma Shepard, but he never prepared breakfast.

From Washington, we traveled to Gettysburg. Not much there except a diorama of the famous battle. I was too young to realize the importance of the site. Next stop was Niagara Falls. Yep, we donned those yellow slickers and rode the Maid of the Mist right up to the base of the Falls. Our last stop was Greenfield Village in Dearborn, Michigan. We walked and walked through the village and toured the Ford Museum.

My special memory was driving by a test field for Ford tractors. There in this field were ten, or maybe twelve gray and red Ford Tractors plowing a field. It struck me as odd to have that many tractors in the same field and in August. Farmers don't plow in August; farmers plow in the spring. Dad said they were testing the tractors and they wouldn't be planting anything this late in the season.

My dad instilled in me the desire to travel. As soon as my sons were old enough to travel, we ventured to many places in our wonderful country. We travelled in a pop-up trailer and eventually in a camper. When our sons went on their separate ways, my wife, Jane, and I traveled to many places in the world. I credit my dad for the desire to see the wonders of the world and how others live.

Modernizing the Farm

FARMING AFTER WWII was very profitable. Several of Dad's friends were building new living quarters. Dad had always dreamed of having a new, ranch-style house. Spreading a house out instead of going up was vogue.

In the summer of 1948, Dad bought up several of the neighbor's feeder pigs. He acquired some 280 head of hogs. He fed the pigs on pastures around our house. When it came time to sell, it took several neighbors to help round up the heavy hogs. The packing plant in Davenport closed their procurement yards for an afternoon just to accommodate his numbers. His check was over $14,000. This was a large amount in 1948, and it combined with the boom in the economy after the war, gave Dad enough money to build his dream house.

Mom, who was always the skeptic, dragged her feet. Her father lost their farm during the Depression and frugality was always her virtue. Dad finally convinced her the money was in the bank.

During the winter months, Dad drew up plans. He figured all he needed was a layout of the rooms and the rest would take care of itself. He contacted the Mueller Brothers in Blue Grass, who said they would build the house. They would need a set of blueprints to work from. Dad hadn't thought of that. He took his plans to an architect. Mom reported the architect looked at Dad's drawings and commented.

"These are very nice, Mr. Bancks. Let me make some suggestions. First, you put your bathroom far away from the bedrooms. I'm sure you do not want to walk through the house after taking a bath. Second, the fireplaces need to be above one another, or you will need two chimneys."

He pointed out several other items. Dad soon realized he needed the architect the Mueller brothers had suggested.

The house construction would take all summer. Where would we live? What would happen to the old house? Our problem was solved by the Van Nice family moving out from Davenport. They had purchased some land to the west of our farm. Mr. Van Nice was going to build a new home someday, but our old house fit his needs. He bought the old house with the agreement we could live in it while our new one was being built. Dad hired some house movers who raised our house and moved it forward from the basement site. They set the house down on strong timbers about fifty feet ahead of the old site. It still gave the construction crew room to build the new house.

We re-hooked the electricity but had no water. Our bathroom in the old house was in the basement. The facilities would not be usable. For the rest of the summer, we used the outhouse next to the garage. We hooked a garden hose to the sink so Mom could have dishwater.

We took our baths in the milk house. I don't know how the others bathed, but I was small enough to fit into one of the washtubs where the milk cans were cleaned. Mom hung little curtains in the milk house windows to provide privacy, even though we lived in the country. I remember carrying my pajamas down to the milk house and wearing them back. As soon as I returned the next person would do the same.

During the construction, I became great friends with Shorty and Ted Mueller, the carpenters. I suppose I was more of a pest than help. I tried to stay out of their way, but it is difficult for a boy nine years old. One day I was playing with my toy cars in the dirt by the unfinished front porch. The house was up, but the floor for the porch had not been poured. The concrete block wall had open holes in it. I mistakenly place my most prized car, a 1948 Studebaker, on the wall. It fell into one of the holes. I reached down to retrieve it only to have it fall further down.

I panicked. My car was gone forever, lost in the wall. I got my mother and through my tears explained where the car had fallen. She brought a flashlight. We could see the bumper. I don't know where Dad was, probably baling hay. Ted Mueller came to my rescue. He made a hook out of an electric cable and by carefully threading the wire down, he hooked the little car's bumper and pulled the toy out. I was so happy I cried tears of joy. I still have the miniature Studebaker.

We survived the summer. The masons bricked the house, and the electricians and plumbers did their jobs. It was getting close to school starting. Mr. Van Nice needed to have his house set on the foundation before frost. He also wanted his sons enrolled at Hazel Dell.

Our new house wasn't finished. The outside was done, but the interior was far from complete. So, we moved from the old house to the basement of the new house. Mom hung curtains from the floor joists to make two bedrooms and a family area. The kitchen area was on the east end of the basement. Mom wanted an extra stove in the basement for canning and butchering chickens.

Everything was downstairs except the refrigerator. We had enough of the kitchen finished to put the refrigerator upstairs, which was a minor inconvenience. The other problem was the piano. It was an old former player piano which weighed several hundred pounds. It was the only item still in the old house. Mary and I practiced as usual, but when the Van Nices were ready for their house, the piano had to be moved.

Dad had a plan. He'd have a piano moving party. He enlisted the card group of friends to help. The Pletts had two young men. Walter Baker had one teenager. Last, there was Lester Chambliss, who was still young and strong.

It was a struggle. First, they loaded the beast on the manure loader and transferred it to the back door of the new house. Everything was fine until they discovered the tall piano wouldn't clear the steel I-beam supporting the floor. They struggled to back the piano up a few steps. Dad removed the step boards. The piano was almost sitting on end by the time it was down. One thing for certain, it was never coming out of the basement, and it never did, until many years later when I cut it up into pieces.

Once in the new house, we spent our first Christmas living in the basement. Dad insisted all the floors and woodwork be hard oakwood. All this wood was too much for him to varnish alone, so he hired a painter from Blue Grass. The painter, an older gentleman, was good, but slow. Also, it was winter, and the varnish was slow in drying. The painter did a good job, for neither woodwork nor the hardwood floors ever needed to be refinished. They've been the same for seventy years.

We always thought we would make it upstairs by Christmas, but Mom realized we would be in the basement until after New Year's. It was close to the day and no Christmas tree. Dad had a solution. He would cut a Christmas tree right from the pasture. It was what his family did when he was a boy.

Our tree was a sticky, dark green cedar, but it was a tree. We used the lights we had. The traditional aluminum icicles were put on the tree but never removed. The tree's sharp needles made the process too difficult. It was near February before we moved upstairs.

The next spring, we planted trees and bushes and established a lawn. It was my job to water the plants every other day unless we had rain. Each plant received fifteen minutes of water.

There were no sidewalks until July. Dad decided to pour them. Instead of mixing and pouring the concrete himself, he ordered the mixture from a concrete company. Just before he got started, he found some steel strips on the road. Evidently, they had slid from a truck. He had a great idea. He would not have straight sidewalks, but curvy ones like you see in city parks. The steel could be used for flexible forms.

Dad called the local lumberyard and told them if someone asked for some steel strips, he had them. The next day a man arrived to claim the steel. Dad asked if he could use them for forms. The man agreed, but he needed them the next week. I think that's why Dad used commercial concrete. We ended up with two S-shaped sidewalks, which Dad thought were very cool.

Our house was Dad's dream. He wanted the basement to be a place to entertain his neighbors. He put a fireplace in the basement for that purpose. He crammed his office into a corner, and salesmen and neighbors met there.

Our first Christmas upstairs was the best. The Blue Grass Cemetery contacted Dad because there were two blue spruce trees on the Thoene lot which were encroaching on the lot next door. They asked if they could remove the trees. Dad called the South Dakota cousins to receive permission.

Logan said, "Go ahead."

Dad cut one down for our house and donated the other tree to the American Legion in Blue Grass.

It was a beautiful but sharp-needled blue spruce tree. Dad thought it was so beautiful he tried to shoot it with the movie camera. He gathered every lamp in the house to provide light, but the film is still a little dark. All that show is the lights on the tree and our smiling faces.

On Christmas Eve, we went to church for the program and candy. When we returned home, we decided to open our gifts early. Dad lit the fireplace and burned some special wood which burned green and blue flames. I sorted gifts.

My big gift this year was a model service station. It had gas pumps that had little hoses to fill the car gas tanks with water. There was a wash rack that pumped water onto the little cars. The water ran down to a basin under the car to be recycled. One end was a ramp-up to the second story. On the other end, you could open the overhead door, drive your car inside, and turn the little crank on top making an elevator raise your car to the second floor. It was fantastic.

Once we moved upstairs, Dad bought an AM-FM radio console with a phonograph player where you could stack several records at once to listen to them. I remember listening to Jack Benny, Amos and Andy, Dagwood and Blondie, and several other shows on the radio. The FM feature gave us the opportunity to listen to KWPC, the Muscatine radio station which was only allowed to broadcast a FM frequency at night. The station used AM during the daylight hours only.

In the evenings, Dad sat by the radio and read the newspaper or his travel magazines. I played with my many toy cars on the rug in front of the fireplace. Mary was in high school, so she had homework. Sometimes she'd find time to play a game of Monopoly. We'd play our game for days, keeping track of who owned what and how much we owed. When it was time for bed, we'd just slide the game to the far side of the room and continue the next evening.

Dad loved music. He'd come home from town with a record album by Wayne King, "The Waltz King," Sammy Kaye, or Bing Crosby. Sometimes he'd buy a single hit of the day. These records were 78 RPM discs and fragile. I was fond of one record, it was "The Old Lamp Lighter" by Red Foley. I knew that song backwards and forwards. I still remember the words.

One day I accidently left the record on the cushion of Dad's chair. When he sat down, Dad broke the record. I was devastated. I cried and cried. Dad must have felt bad because the next week he bought another recording of the song, but it wasn't by Red Foley. It wasn't the same.

I think my father instilled in me his love for music. I've been told in his earlier days he was quite a dancer. Too bad I missed out on that gene. Although he never sang in a choir or chorus, he sang on his tractors while working in the field. In June, Dad would cultivate corn after he finished milking. The daylight was long. We could hear him singing above the purr of the Farmall B tractor. Some of his renderings were of popular songs of the day or of his youth. We could tell when he was getting tired and ready to quit when he started to sing old hymns. When he started singing "In the Garden," we knew he'd be quitting soon.

The old German axiom is "The barn builds the house." This meant the animals always come before the farmer. The animals would pay for the home. Dad decided to remodel the dairy barn to house twenty-two cows. The interior of the barn was designed for cows and horses, but the horses were long gone.

He completely gutted the interior and installed two rows of cow stanchions for twenty-two cows. Each cow had her own drinking cup. In the winter, the cows were housed inside all the time. The only time they were let out was in the morning after milking. The straw and gutter had to be cleaned out and new bedding replaced. Their body heat kept the pipes from freezing.

Dad had already purchased an automatic milking machine. He moved the wash house and attached it to the barn. It became a milk house with running water, a hot water heater, and milk can cooler. It was state of the art for 1948.

The original old barn located just north of the dairy barn needed a new roof and was very out of date. Dad had extra money from building our house, so he decided to build a new barn. It would be identical in size to the dairy barn but used for dry cows and heifers. He hired the same carpenters who built the house. They took the measurements from the other barn and began.

The difference between the barns was the lower part of the new barn was built of concrete blocks. There would be no break outs in this barn. He hired a neighbor with a backhoe to dig the footings. Somewhere in the measurements, the building became two feet longer and two feet wider than the original. It didn't matter, but it made the carpenters work a little harder figuring the rafters.

While the men built the barn, Mary and I were busy pulling the nails from the boards salvaged from the old barn. The boards were to be the sheeting used on the roof. I believe we received ten cents a board. I think Mary did most of the nail pulling. I just stacked the boards.

In September, the barn was finished, all but the concrete lot floor. Instead of hiring a crew to pour the concrete, Dad had neighbors help. Dad's adage of never feel you're better than your neighbors helped. His generosity with his machinery and time had many of his neighbors helping him when he needed help. They mixed and poured several yards of concrete and moved it by wheelbarrow. It was a feat which I marvel at today since the floor still exists although it is quite broken.

The barn was not finished in time to be filled with hay, so it became an indoor basketball court for me for one short winter. It wasn't much of a court. My dad never was into sports, so instead of purchasing a hoop and maybe a backboard, he knew there was an old hoop in a neighbor's ditch.

We dug it out and nailed it to a beam in the haymow. There was no way you could bounce the ball off a backboard to make a basket. Maybe Dad just figured it wasn't worth the effort to build a regular backboard and hoop for one year. Anyhow, he didn't have much invested if my mind changed over the winter.

The next year, the barn was full of hay and straw. Since we did not have a concrete pad in front of our garage like boys in the city, my basketball court was moved outside on the cow lot. Yes, I had a concrete floor for my court, but the cows used it also, therefore; they left their deposits right in the middle. I had to scrape the manure from the area before I could practice.

Only in the summertime when the cows were in the pasture did I have a clean and dry court. It wasn't ideal. In the winter when the cow's droppings froze into chunks, I used the chunks for basketballs. Of course, they didn't bounce, but you had to adjust to the situation. Instead of basketball I dubbed the game "turd ball." Although I always wore gloves, I made sure I washed my hands thoroughly before supper. You learn to play with what you have available.

While the barn was being constructed, the old barn was torn down, which meant there was nowhere to store hay and straw. Dad stacked the hay outside and covered it with several canvases. The straw presented another problem. We had an old hog barn just north of the dairy barn, in poor shape, but the roof was good. There was another small barn where our hired hand lived. The straw would be stored in those two buildings.

The oats were harvested, and straw raked. For some reason, the standard baling crew was one short. Dad needed one more person to pull the loads to the hired hand's place. It was about one mile. Dad asked Ira Dipple if he thought I could drive his Fordson tractor to pull the loads. It was a small tractor and Ira had the wheels set for cultivating, which meant they were spread wide. The little machine looked like a big spider coming down the road.

The decision was made. Nine-year-old Bobby would pull the loads from the field to the barn. There he would exchange hay racks and return. Of course, I was excited. I was going to be an important person and help Dad.

Dad rode with me the first two loads. It took too long because the Fordson was in the low side of its gear drives. Dad shifted the gear set into high drive. There were three gears forward and one reverse.

The third load, I was soloing. I was very careful and cautious. Later in the day, I became more confident. When I was on the road, I shifted into third or high gear. Ira's tractor was a speedy machine when in the high side of the gears. It could travel about fifteen miles per hour.

I pulled out the of field and shifted into high third. I buzzed down the road with a full load of straw bales behind. Now I had to negotiate around a turn at the T road leading to the hired hand's place. What I didn't realize was the load of straw was heavier than the tractor. I cut the throttle back. The tractor barely slowed. I touched both brakes at the same time as I had been instructed. The tires tried to grip the gravel, but they only slid.

The corner was fast approaching. An experienced driver would have driven past the corner and stopped. I didn't. I steered the spider-like Fordson around the curve. The loaded hay rack started to push me toward the ditch. I kept the tractor on the road, but the rack slid sideways. I was on the very edge of the roadbed.

The load started to lean, then the back wheel dipped into the ditch. The entire rack turned over. My tractor was pulled down over the edge, but not into the ditch. I finally stopped in a cloud of dust. The wheels on the front of the rack were on the ground and the back wheels were sticking in the air. It was a twisted mess.

I was shaken. Now what do I do? Somehow, I found a hammer in the toolbox and pounded the drawbar pin out to unhook the overturned rack. I climbed back onto the tractor and slowly headed for the barn where Dad was waiting. I was afraid of what he might say.

I met him at the driveway driving the H Farmall. He looked at me, then looked down the road at the overturned hayrack with its load spilled into the ditch. I started crying. All I remember was him saying, "Are you all right?"

I think he didn't know whether to be angry with me or glad I was not hurt. He turned around and gathered the men. They loaded the unbroken bales on another rack. Unfortunately for me, it was lunchtime. Mom and Mary and the men baling wondered where I went, so Mom loaded everyone in the car and headed for the barn. She soon discovered the accident. I don't know if she realized I was the cause. Anyway, I was finished for the day.

The men were astounded and all of them wondered how I ever got the wagon unhooked, especially the way it was twisted. Once the straw was transferred, the men lifted the rack and turned it back on its wheels. Dad located a bolt to replace the one which broke when the rack overturned. None of the men kidded me at lunch.

Dad got all the straw baled and stored away. He never scolded me. All I remember him saying was, "Next time you'd better start slowing down a lot sooner."

He probably heard more from Mom than I know. I was too young to be driving anything but a bicycle. My guardian angel had to have been watching over me that day. The outcome could have been much worse.

I'm sure it was because Ira had his tractor spread out so wide that it didn't overturn. Dad was impressed with Ira's Fordson. He figured it was a good tractor to teach a son the skills of driving. So, a few years later, he bought an 8N Ford with a mounted two-bottom plow, a two-row cultivator, and a seven-foot sickle mower. It was to be my tractor to maintain. I drove the little Ford for many years. I think Dad wanted me to know how to operate and be responsible for my equipment.

The October Sunday

SUNDAY MORNINGS WERE always hectic. Many times, Dad gave the hired hand Sunday off. One Sunday, Dad either didn't have time to shower in the basement, or if he had, he wasn't thorough. When we were seated in our pew at church, Walter Baker, who sat behind us, began chuckling and whispering to his wife, Maude. He tapped Dad on the shoulder.

"Carl," he said, "did you forget to wash behind your ears this morning?"

Dad reached behind one ear and found a glob of cow manure stuck to his hair. Walt had a good laugh and teased Dad many times. If it wasn't for going to church every Sunday and Dad always being in a hurry, the rest of my life may have been drastically different.

One warm Sunday, October 20, 1951, I was ten years old and getting ready to go to church. My job was to polish the shoes. As usual, mine needed polishing. Mom was fixing something to take to Grandma Shepard's after church. Mary was making beds. We had our radio tuned to the "White Rabbit Bus." It was a children's religious program with singing and Bible stories.

Dad had finished milking and sent John, our hired man, home for the day. It was early and Dad had noticed the gutter on the north side of the barn was plugged with leaves. Dad always found something to fix, but the gutter would only take a few minutes to clean. He carried the big extension ladder from the machine shed and set it against the edge of the roof. The gutter was twenty feet above the concrete which sloped away from the barn.

It was getting late, and Mom started to fret about whether Dad would be coming or not. Many times, he rushed through the shower and shaved on the run. Mom had his church clothes laid out on the bed, as usual, he would be pressed for time. We kids were to stay in the living room or kitchen to be out of his way when he came in. Anxiously Mom went to the back door and called, "Carl, it's getting late."

She heard a noise, but because Mary and I had the radio so loud, she couldn't understand the sound.

"Turn that radio down," she ordered.

It was after we silenced the radio when she heard Dad calling. "Help."

The sound came from the barn. Mom rushed down to the dairy barn and found Dad lying on the floor of the feed room. The feed room door to the lot was open and the extension ladder was lying cock-eyed on the roof of the feed room.

"What happened?" she asked.

"I was cleaning out the barn roof gutter. The ladder slipped. I fell about twenty feet. I tried to call, and the sows thought I was calling them. They came and started to nuzzle me and bite. I had to get inside. Luckily, I left the door open so I could get away from the hogs. I think I broke my leg or hip," he told her. "Get me to the house."

"How am I going to do that? I can't lift you."

"You get Bobby. Have him back the Ford into the barn. I can ride on the drawbar to the front door."

Mom ran back to the house and had me get the little Ford tractor out of the corncrib driveway. My farm boy tractor driving experience gave me the confidence and knowledge to drive the tractor. I drove to the barn and backed down the middle between the cow stanchions to where Dad lay.

The drawbar was only three inches wide. Dad sent Mary to the granary to get a filler board. It fit between the lift arms. Now, Dad had a reasonable seat to ride on. He hoisted himself onto the board seat. I was instructed to drive very slowly while Mary and Mom held Dad's injured leg. I drove to the front door and backed as close to the steps as I could. Leaning on Mom and Mary, Dad dragged himself into the living room and to the couch.

Mom never considered calling for an ambulance. It was just an expensive ride to the hospital, and besides, funeral home hearses were often used as an ambulances. The directors of the funeral parlors used their big vehicles for ambulances to supplement their business. Mom knew Dad had to go to the hospital, and she would take him. She called Wayne Kraft. He drove right over.

Mom said to Mary, "Dad can't go to the hospital in these dirty overalls. Get a clean pair and a clean shirt."

Mary went to find some clean clothes while Mom slid his dirty ones off. Why she thought you couldn't go to the hospital in dirty clothes is beyond me. It was just what you did.

Wayne arrived and helped Dad to our car. In the meantime, Mary called the Van Nices to see if she and I could ride to church with them. We knew after church we could go to Grandma Shepard's. We hurried to redress for church, as Mom drove away. Wayne went to the barn and shut the doors and carried the broken ladder back to the shed. Mary and I waited for the Van Nices to arrive.

It was late afternoon when Mom arrived at Grandma's to pick us up. I was the first to greet her at the door. She told us X-rays showed Dad had broken his hip. There would be no surgery until the swelling went down, probably later in the week. No one seemed alarmed. Broken hips could be mended. Somehow, we would get the corn picked. The Shepard men would take care of that job.

Our hired man's name was John Starkweather. He was a good man and could handle the milking and other chores. The week would be long. The uncles came to our farm twice to help John. They helped him repair the corn cribs to get them ready for harvest.

Mom went to the hospital each day. Visitation hours were from 1 to 3 and 5 to 8. She would go between 1 and 3 so she could be home for Mary and me in the evening.

Because of the swelling in Dad's leg, his surgery to set the broken hip was delayed until Friday morning. Mom left early to go to the hospital. I fed the chickens and my new 4-H calves. Mary washed the breakfast dishes and ran to catch the high school bus at 7:30 at the end of the lane. I hung around home until eight before I rode to school on my bike. Although classes didn't start until nine, it was a warm morning and almost everyone came early to play before school started.

October in Iowa can be cool and wet or warm and dry. Friday, October 26, 1951, was warm, sunny, and dry. At recess, I hurried through my lunch so I could go outside and play. We had a softball game continuing from the morning. I was waiting for my turn at bat when Uncle Vernon's big green 98 Oldsmobile pulled to the side of the road just past the entrance to the schoolhouse.

I said to someone standing near me, "That's my Uncle Vernon. I wonder what he's doing here?"

I watched as Uncle Vernon went inside to speak with Miss Kemper. When he came back, he went to his car. Miss Kemper walked over to me.

I can remember her saying, "Bob, you are to go with your uncle. Come in and get your lunch bucket and jacket."

I followed Miss Kemper and grabbed my things. Then I remembered my bike sitting outside.

"What about my bike?" I asked.

"I'll have someone put it inside the entranceway if you do not come back before school is out. It will be safe," she assured me.

I climbed into the back seat of the big car. To my surprise, Mary was sitting in the front seat.

I asked, "Why are you here?"

She didn't answer. I think she suspected something bad had happened to Dad. I didn't have a clue.

"We're going to the hospital," replied Uncle Vernon.

I was going to see Dad. Great! I chatted all the way to Davenport. The pair in front let me talk.

We arrived at Mercy Hospital and parked on the street. I remember the huge rotunda entrance with stained glass windows which went clear to the ceiling. There were two staircases leading up each side to the registration desk. Because hospital rules were no one under sixteen was allowed beyond the desk, Mary and I had sat many times on the lounges at the bottom of the stairs and waited for Mom while she visited someone.

We reached the top of the stairs. Uncle Vernon spoke to the lady behind the desk. She nodded. Uncle Vernon motioned for us to follow him. I walked beside Mary and gawked at the long halls and all the rooms. Our shoes clicked on the polished terrazzo floor. We took the elevator to another floor and another hallway. I saw Grandpa Shepard poke his head out of a doorway. I began to feel excited. I was going to see my dad.

Mary entered first. I was close behind. Grandpa was standing along the wall. Uncle Jim was sitting on the windowsill across the room. Mom stood at the end of the empty bed. The covers were all clean and white. A pillow was rolled up near the head of the bed. It seemed it was waiting for someone to return. Mom looked worn and distraught.

I asked, "Where's Dad?"

Young boys who idolize their fathers should never have to hear the answer.

"Your father is dead," Mom replied. "He died on the operating table."

Mary burst into tears. She hugged Mom. I'm sure she had suspected something like this. I was dumbfounded and stood silently for a few seconds trying to understand what Mom had just told me. My dad dead? It couldn't be. He and I had big plans. If there was anyone I idolized, it was my dad. I hugged Mom and cried with her.

It was the saddest day of my life. I'll never forget the scene; it's etched in my mind. The bed with the white sheets and pillow all neat and empty. Those walls white and sterile. Uncle Jim with tears streaming down his face. Uncle Vernon pacing back and forth trying to figure out the next step. Grandpa Shepard holding Mom and Mary.

We gathered up Dad's shaver, toothbrush, and clothes. Mom, Mary, and I rode home with Uncle Vernon.

Grandpa drove our car home and Uncle Jim followed. On our way home I said, "Mom, Sandra Jepson said Helen Plett died last night."

Mom was shocked when she heard the news. Helen was a long-time friend of my parents. She and her husband, Ed, had played cards with my parents almost every weekend when they were young.

As we passed the Blue Grass Cemetery, we saw the Pletts's Hudson parked in the drive.

Mom asked Uncle Vernon to stop. We pulled up to Ed's car and rolled down the window. Ed came to the car.

Before she could say a word Ed said, "Did you hear my Helen died last night? She had a stroke. I had to hold her tongue so she wouldn't swallow it on the way to the hospital last night while Don drove. They couldn't save her."

"I heard she died, Ed. I'm so sorry," Mom answered. "I just wanted to tell you Carl died today."

Ed leaned against the car, overcome with emotion. Don came over and helped him away. It was ironic, two great friends died within twelve hours of each other.

It was late afternoon when we returned home. The nice October day had changed into a cold gray windy day. The wind roared out of the northwest, bringing in icy cold air. School had let out. I could see the Jepsons and the Smiths walking just beyond the corner.

We entered the house and Mom suggested I go get my bike. Tomorrow would be a Saturday and the doors to school would be locked. I agreed. I didn't like the idea of my precious bike sitting at school all weekend. Uncle Vernon drove me to school. Miss Kemper was still there, and she helped me lift my bike through the door. Uncle

Vernon told me to start for home as he wanted to talk with Miss Kemper. I was almost to our drive when I caught up with the Jepsons and Smiths.

Sandra stopped and asked, "How come you were picked up by your uncle?"

I didn't know how to reply. The wind was freezing my tears.

I blurted out, "You will find out when you get home."

Sandra and her sisters lived with my Aunt Georgie. I knew Mom would call and tell them the tragic news.

I rode away trying to see through my tears.

Mom's first call when she got home was to Wayne Kraft. He and Dad were very close because my dad was his mentor. Wayne's wife, Valerie, said it was the first and only time she saw Wayne cry. The next call was to Uncle Heinie and Aunt Georgie. They were shocked. Aunt Georgie was already dying from cancer.

Grandpa stayed with us that night. He and Mom talked late into the night. I know because my room was the closest to the living room. I remember seeing the light under my door. Grandpa must have slept on the living room couch.

The morning was busy. John, the hired man, stopped in right after milking. He assured Mom he would stay as long as needed. Our minister arrived at nine to make funeral service arrangements. At eleven we were scheduled to go to Runge's Funeral Home in Davenport.

Uncle Vernon drove us there. We went to an upstairs room full of shiny caskets. We walked around the gilded boxes while the funeral director explained the benefits of each piece. At one casket, the director explained this unit had two mattresses for comfort.

Mom simply looked at him. "That is not necessary. My husband doesn't feel anything now."

Even in her grief she had enough common sense to thwart the Runge sales pitch. We finally settled on a bronze-colored casket with a light tan interior.

"You decide what you want your husband to be buried in. I suggest his suit, if he has one, dress shirt, and tie. We do request underwear also. You will not need to bring his shoes," said the director.

It was difficult to take Dad's clothes from the closet. We knew it would be for the last time. Mary and I picked out his favorite necktie. His diamond stick pin which he always wore was packed. The pin would be removed before the casket was closed for the last time. Every other piece of clothing was a memory. In the end, Mom just sat on her bed while Mary and Grandpa pulled out the different items. The next morning, we delivered his clothes to the funeral home.

Because Helen Plett's visitation was Saturday evening and her funeral was Monday at Runge's, we decided to have Dad's visitation on Tuesday and Wednesday with the funeral service and burial on Thursday. Dad was well known and respected. His visitation would be enormous, and two days would be needed.

We attended Helen's service on Monday morning. It was a precursor for me to get ready for the next few days. I had never attended a funeral before.

Tuesday, we returned to Runge's before three to get ready for the visitation at four. I remember viewing Dad lying there, his hands crossed on his body. The director raised the veil covering Dad. Poor Mom tried not to cry when she touched his hand.

She said, "I would always know your dad because of his bum finger."

Dad's middle finger was disfigured. He caught it in a gear or something years ago. Days later she told me when she touched Dad's hand, she realized he was dead. His hand was so cold. Until then I think she maybe thought this tragic event was just a dream.

Tuesday's visitation was huge, and Wednesday's was even larger. The line of mourners wound through the funeral parlor. The visitation was to be from four to nine but it was almost ten before we left for home on Wednesday.

The Shepards and Bangerts stopped for coffee at our home before going on. It was time to finalize the next day's funeral. On the way home, I rode with Uncle Jim, Aunt Harriet, and my cousin, Jane. My mind was whirling. When I walked out of Runge's that night, I realized I would never see my dad again. I walked in the front door of our house and was greeted by Mom.

I stared up at her and said, "Daddy is not coming home again, is he?"

"Yes, Bobby, your father is never coming home again."

It hit me like a ton of bricks. My dad, my idol, the man who was my mentor, would no longer enter his brick house again. What was my life going to be like now?

I started to cry and then sob. I went into hysterics. Mom tried to console me, but I was too emotional to listen. Finally, Grandpa and Mom took me to my room. They closed the door, but I know everyone could hear my crying. Grandpa told Mom to leave. He would take care of me. Grandpa was known for his stern temperament, but tonight he was gentle and calming. He helped me with my pajamas, coaxed me into bed, and lay down beside me. He stroked my hair, dried my tears with his handkerchief, and softly asked me about my calves and other things. He tried to get my mind changed.

I finally drifted off to sleep. I was exhausted from the events of the last four days. Grandpa stayed all night. He was at the table the next morning. My stern white-haired grandfather was a tender and soft grandfather that night and morning. We had to get ready to bury my father.

Thursday, the day of the funeral, a chilly November wind blew in with another cold front. The temperature was below freezing. It was a far cry from the warm October day just six days ago. In that span of time, my world turned upside down. We arrived at the funeral home an hour before the service. We were given our last private viewing of Dad. I wanted him to open his eyes and get out of that bronze box.

The service was held at Runge's. It was common practice then to have funeral services at the funeral home and not at a church. The mourners filled every room in the building. Several visitors had to stand outside in the cold. People lined the street when the service was over and waited for the procession of cars.

The hearse was first. Mom, Mary, and I rode in the second car, driven by Wayne Kraft. Behind us were Aunt Georgie and Uncle Henie. Behind them were more relatives and friends. The procession was several blocks long. It was a long ride to the Blue Grass Cemetery.

The wind cut across the cemetery like a knife. The tent canvas flapped loudly. Dad's casket was carried to the grave site by his cousins and placed on the webbing. Later Mom wished she had named the neighbors for pallbearers because he was closer to them than his cousins.

Mom, Mary, and I were guided to a row of chairs by the casket. I sat on one side of Mom and Mary on the other. Georgie and Grandma Shepard filled out the seating. Everyone else stood hunkering against the wind. Reverend Rayhill tried to speak in a calm voice, but to be heard he almost had to shout. The wind made the tent flap and groan.

I gazed into the abyss of the open grave. It looked like such a dark place. The walls were carved straight and deep. I tried to understand the gravity of lowering dad to his final resting place. It was good we left before the actual lowering of the casket.

After the ceremony, many friends and relatives drove to our home. Dad always wanted his neighbors and family to enjoy our house. The WSCS women of Sweetland Church provided lunch for everyone in our basement. By four, the only people left were uncles and aunts. Tomorrow I would begin my life without a dad.

We spent the rest of the week at home. Hazel Dell flew the flag at half-staff all week. I missed only four days because the school board dismissed classes on Thursday the day of Dad's funeral. He had served on the Board of Directors for many years.

I know today many young men live without fathers, but in rural Iowa in 1951, fathers were the cornerstone. They were the king. I never had the privilege of having a father to help me through the emotional experiences of my teens. I never had a father to bounce my decisions off when I began farming. I made many mistakes, but they were my mistakes. I had to learn and move forward.

After Dad

MOM HAD MANY ISSUES to wrap up. Dad's estate had to be processed. In 1951, there was an estate tax. All the land, machinery, and livestock had to be evaluated. To defray some of the taxes, the Ford tractor and the small Farmall became the property of Mary and me. Our banker, Ray Tangen, spent hours cutting corners to lower the tax.

The next Sunday evening, Wayne and Valeria Kraft called on us. Wayne asked if he could rent our farm. He knew the land, he knew the machinery, and he had worked for us for several years. I think Wayne had contacted the Shepard uncles before he asked because Mom suspected he would ask her. This was the beginning of a seven-year relationship with the Kraft family.

The plan was for Wayne to take over January 1, but John Starkweather told Mom he didn't feel he could maintain the farm by himself for that long. Wayne agreed to start on December 1. Even at that, it would be almost thirty days before he took over. There was corn yet to be harvested. We had a few hogs not ready for market. Winter was coming.

We stumbled along. John did the best he could. The uncles came often to help. One Saturday the uncles and several of their hired hands helped clean out the hog barn. John didn't have the time to do the job. He had just piled new straw bedding on top of the old. The manure was getting deep. My little Ford hauled several loads from the building and two loads from the chicken house.

The next hurdle we had to cross was harvesting the

corn crop. I know Mom was worried, but she had her brothers to guide her. They, along with Wayne, planned a corn harvesting day. They contacted neighbors and churchmen. This was part of a farming tradition. When a farmer is injured or dies, his crop will never be left in the field. Others will harvest the crop for him or his family.

The day before the corn fest, our neighbors, the Nugents, brought their flight conveyor—we called it an elevator—to fill one of our two corn cribs. We had our own elevator. Wayne had previously mounted Dad's corn picker on the M tractor. He and Ken Nugent opened the fields by picking the outside rows.

The next morning, I wanted to stay home, but Mom insisted Mary and I go to school. By ten o'clock, eleven two-row corn pickers arrived, plus many tractors and wagons. Uncle Vernon managed the crew. I'd say there were seventy-plus acres to harvest. Some men hauled in, some unloaded, some managed the chutes inside the cribs, and the rest drove the pickers. The Farm Bureau Service Company provided free gas for all the tractors. It was a well-organized circus.

Lunch was served in Dad's basement. We had a long table we used for butchering. It was filled with all kinds of hearty food. The neighborhood wives, plus a few from church, served the feast. The *Muscatine Journal* sent a reporter to record the event.

The fields were dwindling by two o'clock. The crew would be finished a little after three. I watched as much as I could from a half mile away until I spotted our car coming down the road. It was Mom. She had changed her mind and figured I should see this spectacle. I got to go home early enough to be a part of the picking party. I wandered around the corn cribs being careful not to get in the way.

At three thirty, the picking was finished. The men lined up all the equipment for a final photo for the newspaper. Although it was not rare for farmers to help each other, this was news because my father was a prominent farmer in Muscatine and Scott Counties. There were eleven two-row corn pickers, eight tractors for pulling loads in, and thirty-two men plus the Farm Service fuel truck. It would be almost dark before some would get home. The corn harvesting day was a success.

The next holiday after the harvest was Thanksgiving and was held at the Bangerts' house because of Aunt Georgie's health. Her cancer was advancing, and she was unable to leave home. I remember her sitting in her kitchen while the other women prepared dinner. Everyone was sad. There was a cloud of gloom over the holiday. I think all realized this would probably be the last Thanksgiving with Dad's cousins. I recall Aunt Georgie crying when we left.

We visited Aunt Georgie the next March. Mom and I brought her a tulip from a florist. She was lying in a bed in their living room. When we appeared, she tried to be strong, but her love for my dad was too much. She broke down and cried. Uncle Henie and Mom tried to console her. Finally, we left for home. I remember her asking me for a hug. I did hug her, but I'll never forget her smell. The drugs and bedding were the cause of her odor. She passed away a few weeks later in April.

December first, Wayne started early in the morning milking the cows. He had some milk cows of his own and he brought them to our place. John stayed until mid-December to help Wayne. By January, John had another job. He could see he would never keep pace with Wayne. Wayne now was farming four hundred acres. It was a huge farm for the time.

In the agreement, the house where John lived would be available for hired men and their families. After John vacated the home, Wayne and the Shepards cleaned the house. John's wife was strange. The couple had several pets. One room was so caked with animal feces the men used a hose and scrapers to clean it. Mom had the house sided with asbestos siding and some new windows were installed. The little house saw many families move in and out. Wayne was a driver. He could out-work two men. The men he hired tried to keep up, but after a few months to a year, they were gone.

December decided to be cold. The temperatures went below zero. Mom dutifully washed the milkers each morning. We had a livestock share rental agreement, so she felt she needed to do her part. Livestock was considered the money makers and not crops like today's agricultural environment. The stress of settling the estate, the corn picking, and getting used to being alone finally got to her physically. She had great pains in her chest and back. Her muscles would spasm. Dr. Schroeder prescribed pain medicine and a harness wrapped around her middle. Mary would help her into the contraption every morning.

One Saturday morning, Mom was feeling very low. Mary and I washed the milkers. When we returned to the house, Mom was feeling a little better. She said she'd had a difficult night. She awoke during the night and absentmindedly put her hand on Dad's pillow. He wasn't there. She figured it was the lowest point of her grieving. Could she go on? She didn't know. I guess you must know your low before you can start climbing the ladder of life again.

In mid-December, Alpha arrived. Finals were over at Ball State where she was a professor. She was Mom's long-time friend from high school and had become a confidant of Dad's. She owned several farms and relied on Dad for advice. She came to be of help.

Although I know she meant well by washing the milkers and helping around the house, one morning before I was up, Mom had had all the help she wanted. There was an argument. Alpha told Mom she'd go to the barn for some fresh milk. The pressure of settling the estate, learning how to handle finances, and dealing with everyone's advice caused Mom to explode. She told Alpha she wasn't in such bad shape she couldn't go to the barn. Mom grabbed the pail we used to collect the milk and threw it at Alpha.

I don't know if Alpha was hit or she ducked, but she picked up the pail and headed for the barn. By the time Mary and I were getting ready for school, Mom was in tears. She knew she shouldn't have exploded. Mom saw Alpha returning. She shooed Mary and me into the living room. We sat quietly trying to listen.

When Alpha entered the kitchen, Mom apologized and apologized. Alpha held her hand up and said, "Edna, it was as much my fault as yours. I was too pushy. Now it is over. We are friends and we will not let this episode ruin our friendship."

It was over. Mary and I went to school. When I got home Mom and Alpha were having coffee and cake at the kitchen table. Alpha never missed a Christmas at our house for many years.

Alpha was a great help through the following years. She helped find Mary a college. She supported Mom by visiting often. She asked Mom to accompany her on many trips and vacations. I think she was one of the many pillars supporting us as a family.

Mom's health improved in January, and with her brothers' help, so did her life. The uncles wanted her to continue with what she would have normally done if Dad was alive. She stayed active in the women's society at Sweetland church and Montpelier Farm Bureau Women.

Sometimes I look back and think maybe she should have gotten a job in Blue Grass or Walcott, but I needed a stay-at-home mom. We continued to traipse to Muscatine on Friday nights for basketball. Mom sat with Uncle Jim and some folks from Sweetland. It was a time when Mom could visit with Uncle Jim about her problems. It was her therapy session.

I would pay my quarter for my ticket to the junior high section. Mary had her student pass in the high school section. I was never alone since my cousin Jane and her friends were always there. It was great entertainment. Sometimes the junior varsity cheerleaders would come to our section and lead us in cheers against the high schoolers. Many times, we would come early enough for the sophomore game. Between games, I'd go out in the hall and purchase my popcorn for a dime.

One time, the varsity players entered the floor for warmups. As I walked along the edge, a ball came right to me. The big high school player motioned for me to throw him the ball. Instead of tossing it to him, I took a shot from the corner. To his surprise, and mine, the ball swooshed through the net. He smiled and gave me the thumbs up. Boy! Was I proud!

After the games, Mom and I walked to the car and headed home. Mary usually stayed in town. There was either a school dance in the grade school gym next door or a dance called "Fun Night" at the YWCA downtown. There, everyone danced to records. Mary had her two cousins, Lucy and Sue, to chum around with. Lucy was old enough to drive, so Mary would stay overnight at their home. The next morning, Mom would pick her up at Uncle Vernon's. Mom always needed to talk to Uncle Vernon anyway.

By February, our lives were changing. All through January, the uncles' hired hands cleaned the chicken house. They used Wayne's manure spreader and my Ford tractor. Some weeks they would be busy and not make the trip. The chicken house would become overburdened, especially when the sun would warm the manure and it would release its eye-watering ammonia.

One Saturday, Mom decided I should try to clean the chicken house myself. As soon as Wayne finished cleaning the dairy barn, I backed the spreader in front of the door. I laid an old window screen in front of the door so the birds wouldn't escape. I started just inside the door with the six-tined manure fork.

I made a game out of the gooey stuff. I pretended I was the U.S. Army, and we were fighting the terrible manure people. The war would be won when the floor was clean. The worst part was scraping the pure stuff from the roosts. It was heavy and very odiferous. I scooped as much as I could with a shovel, then pulled the droppings from the back of the roost area to the front with a garden rake.

In an hour or so, I finished loading the spreader. I put on my heavy winter coat and cap with the ear flaps to head to the field. Spreading the manure was tricky at times. The spreader was ground-driven, which meant the wheels drove the chains connecting the beaters and web. If the ground was slick, the wheels slid instead of turning. I soon discovered if I drove on the cornrow, the stalks helped the wheels find traction. When I returned from the spreading, I parked the tractor and spreader in the corn crib drive.

My work was not finished, though. I had to climb the ladder to the haymow and throw down two bales of straw, lug them to the chicken house, and scatter it around. The hens loved the experience. They'd scratch at the straw hoping to find a kernel of oats or weed seed. They would start cackling like a bunch of ladies at a picnic.

This became my Saturday job for many years. One load of manure was usually enough. There were times the weather didn't permit cleaning so two weeks would go by. Then it would be one load, maybe two. It was hard work and Mom depended on me to do my job. I was her helper, and she didn't want to ask my uncles.

One week, flu was going through our school. I came home sick on Friday night. The next morning, I felt somewhat better. Mom knew the chicken house was overloaded because the week before, the weather was a problem, and I hadn't been able to clean the chicken house. She gave me some aspirin and sent me out. I started slowly. I felt like heck, but there was no alternative. I was the one to clean out the chicken house. It was a man's job. Neither Mom nor Mary was expected to pitch in. Women just didn't do this kind of work.

On that Saturday, Wayne knew I was not feeling well. He sent his nephew, Bob, who was three years older, to help me. He made quick work of the job. We loaded and unloaded two loads of manure before the sun melted the ground and made it slippery. By lunchtime, I was spent, but I had finished the job with help. I put my job ahead of my health.

I found out later in life this is the way it is in agriculture. When you are self-employed and the only person available, you do what is necessary for your livestock. No one would be called to replace you or substitute for you.

Mom did my chores that evening. She thanked Wayne for Bob's help.

Spring finally came, and the soil warmed. It was time to plant the garden. For years, Mom had planted the seeds and tomato plants after Dad, or the hired hand spaded the ground. This became another chore for the man of the house. Mom woke me earlier than usual so I could feed my calves and eat breakfast.

Then, before I went to school, I spaded the garden. I used the same six-tined fork I used in cleaning the chicken house. It was maybe ten inches wide. I would stick the fork into the ground, lift the soil, and turn it over. I would do three widths of the fork in the morning and six widths at night. During the day, Mom planted her seeds.

By the end of April in Wayne's first year, he discovered he had much more land than he and his hired hand could handle. They could plow, but if the dirt dried before a disc or harrow could be used, there would be clods. My little Ford came to the rescue. It had enough power to pull a four-section spike-toothed harrow. Wayne was given permission to use the tractor. It still required some of his time.

One morning he asked Mom, "Do you think maybe Bobby could try driving the Ford and harrow? I'd watch him and make sure I was in the same field."

Mom understood the pressure of planting. She agreed.

That evening she said, "Wayne thinks you could harrow for him after school. Do you think you could?"

I thought, why not? I was already cleaning the hen house. I could drive that tractor anywhere.

The school day dragged slowly. When school let out, I didn't hang around with the boys, I hightailed it home. Wayne was waiting for me. He wanted me to start right out front of the house so Mom could watch. He rode with me for several rounds. He showed me how to use the brake on the inside wheel when I turned and how to overlap just enough to make sure all the ground was harrowed.

By nightfall, I had finished my first fifteen acres. This started my career of harrowing everything from the plowed ground to the disked ground for oats to double harrowing to mix in the alfalfa seed to driving crossways after the corn was planted to cover up the planter tracks. Wayne taught me all the intricacies of fieldwork.

Wayne and Mom made a deal with Miss Kemper. If I kept my schoolwork current, I would be dismissed at noon on fieldwork days. Depending on where the fieldwork was being done, I would ride my bike home or to Wayne's. My Ford would be ready to go. I'd hop on my Ford and drive. I'm sure some of the kids at school envied me for getting out of class, but I had a job to do.

Country schools usually were out by mid-May, so then I started driving full days. I remember becoming sleepy after several hours because of the drone of the engine and the monotony of the field. To keep awake, I'd sing, since my tractor didn't have a radio. Occasionally, I would nod off and drive off course. Eight to ten hours on a tractor is too long for a young boy, but this was only the beginning of my long days.

June was haymaking time. When Dad purchased the Ford, he also bought a sickle mower. It was brand new. Wayne asked if he could use it, and Mom agreed. This model of Ford tractor had an implemented improvement where the operator could raise or lower the implement from the tractor seat with hydraulic lifts. This was a new concept. The very first time Wayne put the mower on the tractor, he raised up the three-point hitch and jammed the power take-off so tight it wouldn't go down. Wayne knew Ira Dipple mowed with his Ford tractor, so he drove to Ira's for advice.

Ira asked, "Did you put on the extensions?"

"No, what extensions?"

"There are extensions for the lift arms. Carl must have put them someplace. I will come up and help you find them."

Sure enough, on the shelf inside the machine shed door were two red ten-inch cast extensions. Ira and Wayne installed the extensions and everything worked fine.

Wayne had purchased Dad's B Farmall with a mower right after my dad bought the Ford. Now he could use two mowers. Bob, his nephew, drove the B. Wayne mowed with my Ford. The first day was fine. He and Bob cut hay for the next three days.

By the fourth day, the hay they had mowed the first day was ready to bale. Someone needed to rake and get the baler ready, yet Wayne still needed to mow hay for the days in the future. This was before the modern mower-conditioners, which allow hay to dry much faster. Hay cut with a sickle mower required three to four days to dry to be safe for storing.

On the fourth day, Wayne asked Mom if she would allow me to try mowing. I'm sure she was reluctant at first. After all, I was only eleven years old. She relented and my mowing career started. Wayne rode with me for several rounds instructing me how to drive, how to brake on the corners, and how to finish a field.

He said, "Most importantly, never, never get off the tractor with the sickle in gear, especially to unplug a slug of hay or mud from a gopher mound."

I could tell by the concern in his deep eyes he was not kidding. His instructions stayed with me all my life. I never once jumped from the mower with it in gear and I still have all my digits.

My summer days were, mowing hay Monday through Wednesday and again on Saturday. In the afternoons we made hay, and I worked the pull-up tractor, which was my Ford. Some of my days were ten hours or longer. I was glad when the second cutting of hay was finished.

The third cutting usually was in September and was small. I was in school, and I wasn't involved in haymaking except on Saturdays.

The oat stubble needed clipping; therefore, I started

mowing again after school. Mowing straw stubble was easy because you seldom plugged up. When the stubble was finished, I started clipping pastures. The Ford and I were good partners.

I was too young for corn picking. I did haul some loads home but never unloaded any. The hoist which lifted the wagons was very dangerous. It was run by a series of cables and powered by connecting tumble shafts spinning unprotected by shields. Everything was exposed. No OSHA laws were in effect. Wayne was not ready to test my skills for corn harvest.

The Krafts

WHEN WAYNE FARMED our place, there were many jobs I couldn't do because of my age. Wayne had the hired help to do the work. The fun part was Wayne and Valeria had four children. Many days during the year Valeria, milked the cows in the evening, and her children tagged along.

Gloria was about four years younger than I was. She had brown hair and a cute little-girl smile. Wayne Jr. was less than two years younger than Gloria. His blonde hair was so curly his mom gave up trying to comb it. Richard, or Dick, another year younger, was the typical farm boy who almost always wore a cap. Linda was very young and seldom was part of our adventures.

The three older kids were my playmates after their chores were done. They fed the cows in the dairy barn and helped carry the chunks of hay. They also had bucket calves to feed. I had the chickens to tend to and my 4-H calves.

There usually was enough time left over for playing. In the winter, we sledded on a hill just west of the barn. There was a small pond just north of the barns. If the ice was thick enough, we'd skate in our boots. None of us had ice skates.

One night in March, we ventured over to the pond. There was ice, but the edge was open. We discovered we could push the sheet of ice with a long stick, so we pushed the slab back and forth. As kids do, we decided if we pushed hard enough, we could ram the slab into the opposite bank. We all found some fallen limbs from the nearby trees and positioned them at the edge of the ice floe. One, two, three, we all gave a mighty shove.

Wayne and Gloria let go. Since my pole was longer, I pushed on. The ice floe hit the other bank. My pole slipped under the ice, and I followed, slipping into the icy water up to my chest.

I turned and tried to crawl out, but the mud was slick. Gloria stuck her stick out and began to pull. Wayne Jr. helped, and I made it back on shore. At first, we were scared, but then we began to laugh. We headed back to the barn.

I was becoming cold as I sloshed my way back with my boots full of water. Gloria and Wayne stayed at the barn while I made a quick exit to our basement. I don't remember whether Mom scolded me or not. Maybe she was just glad I made it out safely. I look back on the moment and think it could have been tragic, but because the three of us didn't panic, nothing terrible happened.

I taught Gloria, Wayne, and Dick how to ride a bike. Our lane was long with a wide grassy strip running alongside. Somehow a bike appeared at the Kraft home, but no one except their dad could ride. He certainly didn't have time for it. Neither did he have time to teach his children. I became the teacher. Fortunately, I had a smaller bike. It fit the trio's legs. The first to try was Gloria. I remembered how my dad taught me. I had Gloria get on my bike while I steadied it.

She asked me, "You won't let me fall, will you?"

"Never," I replied. "Now, I want you to start pedaling while I run along behind with my hand on the fender."

Gloria pushed off. I ran along behind her yelling, "Keep pedaling. Keep pedaling."

I held on for about fifty feet and lightly let go, but I did not stop running alongside.

"Keep pedaling, keep pedaling," I encouraged.

Little did Gloria realize she was balancing the bike by herself.

I quit ordering her to keep pedaling. She continued for another few feet before she stopped. She turned around to see me standing several yards behind.

"I did it! I did it!" she screamed.

"Now ride back," I called.

She stepped over the bike and pushed off, riding on her own. She was smiling ear to ear. I was just as proud as Gloria. I had taught someone a new skill.

In our play, we discovered paint can lids, grease bucket lids, and lids from detergent for the milking equipment could fly in the air and return like boomerangs. Years later, "Whamo" made the same item out of plastic and named them Frisbees. We were way ahead of the times.

One summer the bull thistles were growing thick in the pasture. There were chemicals to control the weed, but we had no way to apply it. When Dad was alive, he and our help would spend days "sticking" thistles or slicing the spiny plant off at the ground. Mom suggested I start sticking the thistles. She sweetened the deal by offering me one cent per thistle. I saw a chance to increase my slim coffers.

When evening chores were done, there was plenty of time to stick thistles. Wayne and Dick tagged along. The thistles were three to four feet tall and sported big purple blooms. Big black and yellow bumble bees love those blooms. I shoved my long-handled spade into the base of the plant. Evidently, Mr. Bumble Bee was disturbed. He flew off and landed on Dick's bare belly. We never wore shirts when it was warm. Dick started to cry. He was sure he was going to be stung. The bee slowly crawled up his chest.

I yelled, "Don't move. Maybe he will fly away!"

Wayne and I waited and watched as the bee moved upward. Dick was scared but he didn't move. Finally, I reached out with the tip of my shovel and gently touched the bee's behind. It was enough. The bee flew away.

We all were relieved. I was glad my coaching and Dick's listening resulted in no bee stings. That also was the last time I stuck thistles without a shirt, but I didn't stop sticking them. Even today I hate those thistles. I don't stick them anymore. I ride my UTV and spray chemicals on them.

Although it was the Kraft kids I played with, it was Wayne Senior who taught me the practical elements of farming, things my dad didn't live long enough to teach me. My dads were Uncle Vernon, Uncle Jim, Grandpa Shepard, and Wayne Kraft. The uncles helped with 4-H and management, but it was Wayne who showed me the basics. I think he felt because Dad helped him get started, he should help me do the same. I learned his way of building fences, how to cultivate corn, mow hay, how to load bales on a hayrack and to work hard. He was a mentor to me.

Wayne was a tall, thin guy. He was the type who probably could eat five meals a day and not get fat. He grew up in a family of fourteen siblings, so he was accustomed to hand-me-downs. He never went to high school but was sharp as a tack. I was always amazed by his ability to work figures in his head. He most always wore a smile. His mojo was the unique names he had for certain objects in his life. A pocket watch was a turn-up, a hammer was a tacker, gloves were hand shoes, his old F-20 tractor was Betsy, the Farmall B was the Puddle Jumper, and so on.

He had his own philosophy on building a fence, like placing the braces horizontally instead of at an angle. You always put more dirt back in a post hole than you took out. The rule was if there are two people building a fence, the person who digs the hole does not tamp the dirt back in.

He had baling hay down to a science. When he started farming our place, he bought a share of Dad's New Holland baler. It was a monster of a machine. The baler had its own engine. Tractors at this time had no ability to independently transfer engine power to the trailing implement. In today's world, farmers call this feature Independent Power Take-off. This means the tractor operator can operate the power supply from the tractor without disengaging the clutch for the entire tractor thus stopping or slowing the tractor itself. When tractors were engineered to operate with remote power equipment, extra engines on pulled implements became rare.

This old baler was powered by an air-cooled Wisconsin V-4 engine. It ran well until it became hot. If for some reason it stopped, it was very difficult to start again. This was very frustrating. The baler also had a throw-out clutch on the hay pick-up so when the driver came to a large bunch of hay, he could control the intake. He did so with a rope tied to the tractor seat. The driver could stop the tractor and ease into the hay pile, and if the pick-up gathered too much, the driver would pull the rope and stop the hay. It saved shearing safety bolts.

Wayne had a lot of hay to make. We had two barns to fill, and he had one huge barn at his home. We made hay almost all summer. One warm July day I was moved from driving pull-up to pulling loads in from the field at Wayne's.

The Kraft farm was long and narrow. The hay fields on the south end were some fifteen to twenty minutes of driving time away. This day when I arrived in the field, I found Wayne had already unhooked the full wagon. I whipped in behind the baler and hopped down to unhook my empty. Wayne's hired man, Jim Marberry, was sitting in the shade of the rack by the rear wheel and he looked beat.

Wayne approached me and asked, "Do you think you could drive the baler? Jim is over-heated and needs to rest. I'll send him with the load, and I will load and you will drive."

Of course, I said yes. I may have been thirteen and still had some growing to do, but I was tall enough to reach the clutch. For the next two summers, Wayne and I baled hay. I drove while he racked the bales.

On the baler was a throw-out lever which stopped the hay pickup. It was a rope attached to the back of the tractor seat from the baler. The purpose of the throw-out lever was to stop the hay from feeding into the baler too fast. Occasionally, I wasn't quick enough to stop the hay pickup and the baler would shear a safety pin. We'd pull out the stuffed hay, re-insert a new bolt, and hope when Wayne threw the engagement lever the baler would shove the plug through.

Wayne instructed me to drive with the outside on the hay pickup edge of the row. He claimed it made the hay flow better. I tried my best. Wayne never chewed me out—after all, I was the landlord's son—but I received some stern advice. I remember Wayne was like my father. You listened and heeded his instructions. You didn't want to be told twice.

One day we were baling at Wayne's, the weather was ideal, and the baler punched out bales four or five a minute. Wayne hired some teenagers to help load. I guided the old M around the field. I don't have any idea how many bales we tied that day, but by seven I was worn out. My arms and legs ached. I drove my Ford home and parked it in the shed. Mom came out to greet me.

"I fed your calves," she said.

I answered, "Thanks," and almost collapsed on the terrace by the house. I was so tired I cried.

She helped me up and told me to take a shower. She'd have supper ready when I finished.

While I was eating, she said, "You remember Mary is having a pajama party tonight."

At first, I was happy. It would be a fun night. I knew all her friends. But I have heard of runners hitting a wall when exhausted. After supper, I hit my wall. I felt sick. I had a pounding headache. My body felt like a log. Mom put me to bed early.

Then Mary and her friends decided to sit on the front porch and laugh and giggle. My room was next to the porch and my windows were open. Finally, Mom asked the girls if they could go elsewhere to talk. A couple of the girls came to my window and said they hoped I felt better in the morning. I did feel better the next morning, but I had no energy. Mom talked to Wayne about maybe not having me drive so long. He admitted it was a long day and probably wouldn't happen again. He did appreciate my driving, though.

A few days later we were baling a field we called The East Hills. They were a series of short sharp and steep hills. I drove the hills with skill, always watching Wayne on the rack. It was late in the afternoon and the only windrow left was the outside round. These were most always baled last because they contained two swaths of alfalfa. It took longer for the bigger row to dry.

I started around in second gear. The baler pounded out the bales. We were almost finished except for just down the hill and up the draw. I put the M in low gear. The weight of the baler plus a three-quarters load of bales on the rack began pushing the tractor forward. The tread on the tires was worn and traction was not good on the hard, sun-dried soil. I pushed as hard as I could on the brakes. The hard ground refused to slow us. I made a young driver's mistake; I pushed in the clutch.

I heard Wayne yell, "Don't put the clutch in."

It was too late. The tractor, baler, and load all started to gain momentum. At the bottom was a ditch, and it was approaching fast. I let the clutch back and cranked the steering wheel to the left. I looked back at Wayne. I was scared.

Near the bottom, the tires found some soft dirt and everything slowed with a jerk. Wayne jumped from the rack and ran to the tractor. He leapt onto the seat behind me and jammed on the brakes. The baler slid sideways, followed by the hayrack. The tongue of the rack bent and most of the bales tumbled off. We came to a stop with the bale chute up on the rack, but we were not in the ditch. Wayne sat for a moment. He probably would have liked to give me some choice words, but I think he realized I was only thirteen and had made one of many mistakes. You learn from your mistakes.

Instead, he said, "Bobby, never put the clutch in when you are going downhill. Now you must pull ahead while I attempt to unhook the rack."

By the time we unhooked, the men from the barn arrived. We reloaded the bales. Wayne drove for the last 500 feet. I never forgot my lesson. Never again did I throw the clutch in, going downhill.

The next year, my driving improved. It was the old '77 baler that gave Wayne fits. It wasn't an electric start but a hand crank engine. Most days it started with one or two cranks when it was cool. We never shut it off once it was started. Wayne would let the baler idle during lunch.

One day, while we were having a lunch break, the fool motor popped and sputtered, then died. Wayne cussed a bit, but we had just started to eat, so he figured it would be cool enough to start later. When it was time to return to baling, Wayne gave the engine a mighty crank. It popped once. For the next fifteen minutes or so he cranked and cussed. I'd only seen Wayne madder once before when he fired a hired hand.

Finally, he said, "I'll show this S.O.B. I'll hook the pulley to the tractor pulley." One of the neighbors thought this might be dangerous because he had to remove the drive belts, and how would he replace them? Wayne answered, "I'll figure that out when I get this dammed thing started."

He turned the M around and aligned the belts. He threw the pulley in gear. The engine had to turn over. Wayne revved the tractor motor. The baler began to fire. It popped and snorted, but it started. He moved the M ahead to loosen the belts. The next job was replacing the belts from the engine to the baler flywheel. There were multiple belts. I don't recall how, but Wayne slipped those belts on without losing or pinching a finger.

The very next day, after milking, Wayne came to the house.

"Edna," he said, "would it be okay with you if I used the baler for a trade-in on a new machine?"

Mom knew about the trials of the previous day. The only reason Wayne had to ask her permission was because she still owned a one-half interest in the baler.

"Of course, Wayne. I have no use for the baler, and if it will help you, go ahead," she replied.

The next day a new Holland baler was delivered. It had a much smaller engine, and the whole baler was smaller and more efficient. He timed the plunger at sixty strokes per minute. It took ten to twelve strokes to a bale. If the windrow was even and the driver was alert, five to six bales per minute could be tied. That number of bales normally required two men on the rack. A rack held 72 bales. Every twelve to fifteen minutes, there would be a load. Then it was time to change wagons. On a good day, starting at one o'clock and going to six, we would bale fourteen loads.

When I was old and big enough to load bales, Wayne taught me how to load those square bales. The knots were always on the top of the bale. As the bale exited the chute, they tended to bend slightly. The rule was you always stacked the bales with the knots facing down. If one packed the bales tight enough together and knots down, they were almost impossible to shake loose.

It was late afternoon; we were baling straw. I was driving baler. Bob Strong, Wayne's nephew, and someone else was loading. The rack was becoming full, and we had several rounds to go.

The guys on the rack decided to stack the straw higher. The field was fairly level. Soon there were over ninety bales, and we had one more round to go. Bob gave me the thumbs up to keep going. I idled back. When we wrapped up the last bale there was 125 bales on a rack which usually held 72. When we pulled into the farmyard, Valeria took a photo of the load. We were two young men who knew Wayne wanted to finish baling straw and we were sort of showing off. It was good the field was level, or we would have never accomplished the load.

I drove the pull-up for several years. It required skill. Every barn was different. A very long rope was used to lift the hay bales. Each rack contained eight or nine lifts. The pull-up driver had to listen for the man sticking the bales. He would yell, "Go ahead!"

At that point I would start backing up, tightening the rope, and lifting the load up to the long T-shaped track running the length of the barn. The load would click in and glide along the track until the men in the mow hollered, "Trip it!" By this time, I was able to see the man on the rack pull a smaller trip rope signaling me to stop. I would then change direction and return to my original position, stopping about two-thirds back. The man on the rack would pull the carrier and the sticking forks back until they were released and began to drop down.

This was the tricky point. The pull-up driver must wait until the forks are descending before slowly moving forward. If I didn't stop, the forks would plummet down and maybe injure the man on the rack.

All this was wearing the inch-thick hay rope. In time the fibers, which made the rope, became weak and needed to be replaced. The rope couldn't be repaired; it could not be knotted, or it would not clear the several pulleys it threaded through. Now Wayne had a huge barn at his place, and the rope needed to be replaced. He purchased two hundred feet of rope. We installed the new rope in the morning before baling.

On the second fork load up, the rope began to twist. The bales spun around several times. Bob Strong was the stick man. He tried to unspin the load. I could not see what was going on as I retreated. The bales reached about halfway up the barn before Bob jumped from the rack and yelled at me to stop.

"The rope is twisted. I think if you come forward, I can spin them back," he said.

I moved forward. The rope refused to untwist or return the bales to the rack. Bob tripped the forks and all the bales fell on the rack. It was evident we needed Wayne. Someone drove out to the field and brought everyone in. Wayne assessed the situation.

The hay inside the barn was fifteen feet from the door sill where the hay entered. The hay forks were suspended thirty feet above the rack. While the rest of us watched, Wayne climbed the wall to the door sill, a wooden six-by-six lying across the bottom of the door, making a five-and-a-half-inch surface. Wayne stepped out on the sill, grabbed the twisted rope, and began to untwist it. Bob pulled the forks gently down. We all were amazed to see Wayne walking out on a six-inch piece of wood, thirty feet above the ground, and calmly doing the job.

When he got down from his perch one of us asked, "How did you do that?"

His answer, typical of him, was, "You can walk on a two-by-six on the ground and not fall, so why can't you do it thirty feet in the air?"

He was a marvel and a daredevil. Most important was he had a job to do. He was the leader. He taught me sometimes you can do risky jobs if you are smart and not afraid. For the next several loads, Bob and I were very careful to keep the rope from twisting. Eventually, the weight of the bales stretched the rope, and this was not a problem any longer.

The mid-50s were good years. Wayne had the new baler. Next he purchased a new, bright orange Allis-Chalmers WD-45 tractor. It was one of the first tractors to have live power take-off capabilities. The era of long drive belts was passing. Power shafts for transferring tractor engine power to an implement became the norm.

This WD-45 also had hydraulic power. To raise the cultivator, you just pulled a lever. My 8N Ford had a cultivator, but the mower was generally mounted on it.

The WD-45's job was cultivating corn. Cultivating corn was a yearly task. It is an implement holding V-shaped shovels which when mounted on a tractor to stir the soil and root out weeds growing between the rows of corn.

We tried to cultivate the corn field twice each season before it became too tall to drive through. In rainy years, once through was all we could find time for. Guess who got the job of cultivating? Me! I'd mow hay in the morning and cultivate corn all afternoon. Some days it was hot riding that orange rig.

One day I was all by myself up north of the creek at our home. Wayne and crew were at his place. I had my jug of water hanging from the seat. I figured no one was there but me and some deer, so I removed my jeans and drove in my underwear. It was much cooler. My snow-white legs soon became red. I didn't realize it until too late. I suffered that night, but I was too embarrassed to tell Mom.

The cultivator was a two-row. Wayne planted with a four-row planter. If you started right, it made little difference. Wayne told me, if you looked across the hills of corn, you wouldn't get on the wrong row. This was the time in farming that you planted three kernels of corn in one spot. When farmers cultivated with horses, they would cultivate one way, then cultivate 90-degrees difference. It made little hills around the three corn plants. If you cultivated rows two and three, you were okay. That left rows one and four to go. To compensate, and if Wayne drove real straight when he planted, you could do rows one and four without plowing out corn.

This was rare because of our hilly land. The field west of the house was oddly shaped; rows never came out perfect. Wayne started on the southside for most of the field, but when he got almost to the north end, he started again from the north side. The rows met in the middle.

I had cultivated all day. This last four acres, I forgot to start on the north side. Of course, I worked the wrong rows two and three. I was finished when I realized what I had done. This meant doing it all over. I started on my task of covering up my mistake, but the sun was going down. I was bone-tired and still had chores to do. I had to quit and admit my mistake to Wayne. This was the most difficult task. He didn't get upset; he just told me to go back and finish in the morning. That night it rained. The corn grew and I never got back to finish the field.

As all farmers know, along with the good years, there are bad years or other unforeseen events. Dairy farmers once used live bulls to breed their cows. Having bulls around was a necessary evil. Some Holstein bulls were of good temperament. They could be trusted to run with the cows and do their thing. Then there were times when we didn't want the cows bred, so our poor bull was housed in a small pen on the north side of the loafing barn. It was well built out of 2x12s and 2x6s.

One morning before I was heading to school, Mom and I heard a loud crash at the loafing barn. I saw Wayne running after the bull. By the time I got outside, Wayne was cranking up Ol' Betsy. It was early, so I ran down to see if I could be of assistance. When I reached the barn lot, he yelled at me, "Open the gate, then you get back to the house."

I opened the gate and he drove through. The overhead door had been smashed to smithereens. The bull was tearing around the lot, bellowing and pawing the concrete. I ran back to the house as ordered. Wayne started to drive the mad animal using Ol' Betsy. Mom came out and told me it was time to go to school, and she'd help Wayne.

When I returned that afternoon, the bull was corralled. One of the neighbors brought over a head harness with steel eye coverings. The bull could only look down. Mom said it was quite an ordeal putting it on. The bull went berserk. It took several men to lasso and tie him up.

Next Monday, the bull was loaded for the sale barn. I'm sure he became a baloney bull and ended up on someone's sandwich.

Mom and Wayne discussed what to do next. Wayne was a progressive dairyman and he'd read about artificial insemination, or AI as he called it. I had no knowledge of what they were talking about, so I asked, "What is AI?"

Nobody would answer me. Finally, Mom told me to go check the chickens.

Later at supper Mom scolded me. "Why did you keep asking about AI?"

"Well, what is it? Is it a secret or something?"

She explained the process. I don't know why it was such a secret. I knew it took bulls and boars to make calves and pigs and I didn't learn that from a sex education class. It proved a good thing that the bull went crazy. Not only was AI safer, but it also gave Wayne access to better bulls. The herd improved greatly.

Wayne had the herd producing at a high level. One of the aged cows which had been part of Dad's original herd was ready to be replaced. She was sold. When she was slaughtered, the kill line discovered her lungs were full of tuberculous. The State Veterinary called and told Wayne the entire herd must be tested. The veterinary came the next day and injected each cow under her tailhead. Three days later he came back and read the injection site. If it was swollen, the cow was infected.

It turned out that all but one cow was infected. The entire herd of milk cows had to be destroyed. Tuberculous is a contagious disease. The government had an eradication program to rid the nation of this dreaded disease. This meant a great loss to both Wayne and Mom since we were on a livestock share lease.

Fortunately, the outside heifers tested negative, as did my 4-H calves. The heifers were immediately moved to Wayne's place. My calves went to the hired man's place. I had to drive a mile each morning and night to feed them.

The cows were transported to a special slaughtering plant in Wisconsin. Wayne and Valeria watched the cows slaughtered. Not one passed inspection, even for pet food. The meat was cooked and ground for meat meal. The government gave us some compensation, but it was far short of the value of the animals.

The next decision was, do we start all over again or quit? Dairying, although a demanding enterprise, does provide a steady income. Wayne and Mom decided to restock the herd. It was a financial burden for us and for Wayne.

The next step was disinfecting the dairy barn. The State ruled we must be out of dairying for three months. In that time period, we washed and scrubbed down the dairy barn. All equipment had to be sterilized. Any part of the milking machine that touched the animal had to be replaced. Sunlight and rain would take care of the outside.

Wayne hung ultraviolet lights inside to kill the tuberculous spores, but he didn't realize direct exposure to ultraviolet light is harmful to eyes. He worked in the barn one afternoon, and that evening, his eyes began to burn. He told us later he had never experienced so much pain. He thought he was going blind. For several days, his nephew Bob and I handled most of the chores. In time he recovered.

The State Veterinary inspected our facilities and declared them safe to repopulate. Wayne found a dairyman who was going to retire, and he sold his entire herd to us. We were back in business.

We always wondered where the TB came from. When we backtracked, Mom remembered Dad going to a herd dispersal sale. The dairyman was hit with a tuberculous infection. The cows left had supposedly tested negative. Dad bought one of those animals. She was probably the carrier, and over the years, she infected the rest of the herd.

In addition to the cows, our whole family, plus Wayne's, had to be tested for tuberculous. Fortunately, we all tested negative. We were encouraged to pasteurize our milk instead of drinking raw milk as we had for years.

It was just one of the lows of farming. We had to start climbing up again. Later in my years of farming I had many lows from droughts, hailstorms, disease, and just plain ignorance. I found I had to shake my head and recover. I never gave up. There was always next year.

When we purchased an entire herd from a retiring dairyman and moved to AI for breeding, the herd milk production increased dramatically. The offspring, if they were heifers, became more valuable. One mid-July evening, one of the cows was due to calve at any time. That night, she didn't return to the lot with the other cows. Nature's way with cows is to have your baby away from the herd. Maybe she was trying to hide the young calf from wild animals.

Wayne sent his children out for a search. This was now their chore and not mine. They found her across the creek with a calf. Wayne was busy milking, so he decided to bring her in afterwards.

At seven, clouds began to gather. It looked like rain. Mom and I were eating supper when Wayne knocked on the door.

He said, "I have a cow and her calf across the creek. I need to bring her in before the rain. I figure if Bobby drove Ol' Betsy and I held the calf on the platform, we could bring them in."

I was thirteen and could barely reach the clutch and brake pedals on the F-20, but I couldn't turn Wayne down. We drove to the creek and found the cow. Wayne gathered up the calf and held it. He told me to put the tractor in low gear. The mother cow followed obediently across the creek.

But when we started up the hill, she changed her mind. She turned around and trotted back to the calving site. We followed. The lightning flashed, followed by a rumble of thunder. Rain was not far away. It was becoming dark, and the F-20 had no headlights. Time was running short.

This time, as we reached the hill, the cow stopped again. Wayne jumped off and laid the calf on the ground. Mother cow inspected it. She mooed to it. The rain started. Wayne picked up the calf and wrapped it around his shoulders.

"You go on home with the tractor. I'll walk the other way to the barn with the calf. I hope she follows me. If you get to the barn before me, open the gate."

"Okay," I answered.

The rain increased. In minutes it was pouring. The path back to the house was lit only by flashes of lightning. There was one other source of light. Ol' Betsy was wearing out, and to keep her running, Wayne had set the carburetor to a very rich mixture of gasoline. In the dark, a blue flame shot out of the top of the exhaust pipe, which was a drainpipe from some roof.

I crept up the hill, soaked. When I came to the pasture gate, I had to get off the tractor and open it, drive through, and close it. Wayne and the cow were already in the barn when I arrived. My path back was much longer than his and the cow's. The water was dripping from me as I went to the barn where Wayne was. He was as wet as I was, but he had a big grin on his face. It was a heifer calf. Our effort for the night was rewarded. The calf would be a replacement heifer for sure. It was another lesson of putting your livestock ahead of your comfort.

After six years of working our farm, Wayne was tired. Hired men who could work at his pace didn't exist. His nephew, Bob Strong, who lived with the Krafts, graduated from high school and was leaving. Bob was a lot like his uncle, long, lean, and lanky and one heck of a worker. Wayne decided to farm just his own place. It was a shock to us. I still remember him standing in the doorway of the kitchen telling Mom and me he was quitting.

It would be quite a change. I had one year left of high school, and my grandfather and uncles were determined I attend college.

Wayne Kraft never really left. He was always checking on me years later. We continued to bale hay together and traded machinery. One time I was replacing a line fence east of the home place. Somehow, Wayne found a reason to check on me. I was just finishing the corner posts when he arrived.

He said, "Looks like you got a good start. I see you put the braces in right."

I looked at him and thought, Wayne, you just couldn't stand it. You had to make sure I was doing it right.

I was proud he had approved my work. In later years, when my sons were big enough to help, Wayne and I baled less hay together. He always remained a good neighbor. The day Mom died, I said to my wife, Jane, "I must call Wayne Kraft right away. He and Mom were close."

Wayne wasn't shocked, since he knew Mom was in the hospital, but he was saddened. I visited with him for a few minutes. Later, I talked with him and Valeria at the wake after Mom's funeral. There were many memories recalled that day. I had learned much from this man. Our years farming together and him teaching me the fine points of farming you don't find in textbooks. I knew my mother thought highly of Wayne since he was the first person my mother called when Dad died. He was a special person.

Several years later, Wayne Jr. called me to tell me that Wayne had passed from a cerebral stroke. Now all I have are memories of an interesting man who began our relationship one evening in July beside an Allis-Chalmers combine when my dad was alive. He guided me through my growing up years, putting up with my mistakes, and following me through my first years of farming. He was my second dad.

My 4-H Projects

JUST BEFORE DAD DIED, I joined the Montpelier Busy Boys 4-H club. Dad and I attended the annual new member meeting in October. Jack Van Nice and I were the only two new members. The meeting was held at Patterson School. We had a hot dog roast and apple juice pressed with apples from Lyall Paul's orchard.

We were all asked what kind of a project we were going to have for the coming year. I looked at Dad, who answered, "Bob's going to have two baby beeves."

I had an idea that was to be my project but at ten years old your father is really in control. He was the one who had to buy the animals, provide the feed, and take care of the housing.

Most of the members had beef projects. A couple had hogs and sheep. This was 1951, so all boy 4-H members had livestock projects. A week later, Dad and Mom drove Dad's new gray International pickup to the Thoene Ranch in South Dakota and purchased two Hereford calves, which became my project.

One had a head covered with white curls. His name would be Curly. The other calf was shorter, and rounder in frame. His name became Whitey. Dad made a special pen in one end of the barn. It was my responsibility to care for those calves. At first, they were so skittish I couldn't touch them, but in time I could pet them and curry their hair.

Taking care of large animals was a new experience for me. I was just a dog and cat man until now. Little did I realize at the time soon I would be raising my calves without Dad.

Every morning before school, I carried water and feed to the pair of calves. I'd take my five-gallon bucket to the water hydrant on the south side of the barn. I'd fill the pail and dump it into the calves' watering tank inside.

Opposite the calf pen was a room Dad had constructed for the purpose of storing feed for the dry cows or later beef animals. I was allowed to use the feed room in the barn to store my feedstuffs for my calves. I had no knowledge of feeding cattle, so Uncle Jim sent his feed salesman from Kent Feeds, Russ Brannen, to help me.

Russ showed me what to feed the calves, how to mix the various ingredients, and how much to give them at each feeding. Russ built me a mixing box for the ratios. It didn't take me long to figure out to mix my morning feed in the evening, so it was ready for the next morning. The feed was a crazy mixture of corncob meal, rolled barley, a protein called Baby Beef, and a mixture of bran and molasses. The calves also received some hay. I tried to feed my calves better hay than the dry lot heifers, but the heifers liked the hay also. They could stretch their necks around the stanchions and grab a bite of my hay with their long tongues.

One day someone either left the door unlocked or a cow rubbed the door latch, and the dry cows got into the feed alley and inside my feed room. It wouldn't have been bad except I had no door or gate on my feed room. The heifers tromped inside, ripped open the feed sacks with their sharp hooves, ate a good part of my corn cob meal supply and, of course, pooped all over everything. It was just one of the accidents happening with livestock.

I repaired my mixing box the best an eleven-year-old boy could do. I found some unused boards and two rusty hinges and built a gate to close the feed room. Without my father's guidance, I sawed and nailed the gate together. It was just one of those learning experiences I did myself. I must have done a fairly good job, since the gate is still there today.

It was a very cold December and January that year. The water would freeze if the calves didn't drink it all. The snow drifted over my path to the hydrant. On very cold mornings, I had to lift the water pail the calves drank from and take it outside to beat the ice out. My hands were stiff by the time I finished chores. Once I got back to the house, I'd sit in front of the heat register to thaw out.

I never had Wayne feed my calves when it was cold or rainy. He did grind some corn cob meal for me and back the wagon close to the feed room. I shoveled it from the wagon. Mom fed the calves a few times when I was ill, but generally, I did all the work. I know some of my peers had their fathers take care of their animals on bad weather days, but it wasn't a luxury I had.

The calves, although penned in a small pen in the back of the barn, were skittish. They were used to big open country. I tried to pet them, but they ran away from me. I wasn't big enough to corner one of them or even put a halter on one. Whitey was the tamest. By February, I could pet and curry him. Curly would not let me touch him.

One Sunday in early March, the Shepards were at our house for dinner. Uncle Jim surprised me with two new show halters and two tie-up halters for my calves. He had delivered some cattle to Chicago and purchased the equipment at the stockyards. After dinner, Uncle Jim suggested we start to break the calves to lead. He had brought some coveralls and boots. Uncle Vernon had brought boots also. We went to the barn to halter the animals. Of course, they were not happy with the process. The uncles climbed into the pen and roped the calves, then tied them to posts.

"Go get your Ford tractor," said Uncle Jim, "and bring it around to the gate."

I ran to get the little gray machine. By the time I drove to the barn, the uncles had the stubborn mavericks to the gate. We tied their halters to the drawbar and slowly started down the lane. The calves planted their hooves and refused to walk, but the tractor didn't budge. They had to follow or be dragged. The uncles walked behind and pushed sometimes.

When we returned them to the barn, instead of letting them lose, we tied them to the posts again. The leather halters stayed on their heads. Uncle Jim also bought me some cotton rope leaders with hooks on the end. Now all I had to do was to hook the ropes to the halters while Curly and Whitey ate and secure them to a post. Soon they were calm and seemed to enjoy my currying.

The breaking to lead would be a bi-weekly occurrence for me and my 4-H project until May. In May, the uncles became busier with field work and the calves seemed to enjoy getting out of the dark pen, so I tied the calves to the tractor and led them myself.

The March escapades were an eye opener for my uncles. They realized the difficulty I had doing chores. Wayne also had a difficult time keeping the water tank open for the outside cows. To my surprise, one day when

I came home from school, Uncle Vernon and his men were digging a trench to lay a pipe for a hydrant and an automatic cattle waterer for my calves and Wayne's dry cows.

Because all the digging was by hand and the pipes had to go through a concrete block wall up through a concrete floor, the waterer was placed just inside the wall of the building. At first, I was elated at the prospect of not carrying water anymore, but one problem became evident quickly.

The dry cows and heifers fed at the hay bunk in the middle of the barn. The water was at their back ends. Many times, they would back up into the waterer and deposit manure in the water. My calves did it a few times, but the cows did it almost every day. Of course, I was the person who got the job of cleaning out the filth. As soon as the cows were let out to pasture, my job became less strenuous.

Mom also received a pipeline to the chicken house. There would be no more frozen water for the laying hens.

By June, the chore was to give the calves a bath. Did they need it? I think it did calm them and make me work harder. As the fair approached, the effort was more appealing. Uncle Jim bought me a curry comb, a liner, and a stiff brush to enhance the hair on each animal. Curly and Whitey were range cattle and not the low-slung, short, thick calves the judges were looking for.

Uncle Jim was my mentor when it came to the 4-H calves. One Sunday afternoon, he lined me up for a grooming class at the Kent Feed Experimental Farm. It was taught by the same crusty cattleman, Russ Brannen. He was an expert on cattle and had a sense of humor which appealed to young people. We hauled the two Herefords to the Kent Feed farm. Russ showed me how to clip the curly hair from Curly's head. We clipped the tails right down to the long bushy hairs on the end. The tousle then had to be picked apart to make it as big as possible. This was to try and cover the space between the calves' legs, of which my calves had plenty to cover.

Another trick for showing a calf is to make them stand squarely on all four legs. You did this by using a long stick with a sharp tack on the end. You would poke the stick at the animal's hooves until he stood squarely. Again, Uncle Jim came to my rescue. He purchased a wooden show stick when he was in Chicago. It worked fine.

In mid-August the West Liberty Fair started. All the 4-Hers hauled their prized projects to the fair. If you had livestock and lived on the other side of the county, like I did, the boys stayed at the fairgrounds overnight. This truly was a new experience for me. My only staying away from home had been my many overnighters at Cousin Jane's.

Luckily, my neighbor, Jack, was doing the same. Our parents solved the dilemma of where to stay. One of the older boys in our 4-H club, Mark Kemper, had a tent. I don't know if he was pleased or not, but he became our big brother for the five days of the fair. Jack and I slept and ate breakfast with Mark.

In the morning, around six, we fed our calves, cleaned their stalls, and bedded them down for the day. We finished in an hour, met Mark, and walked down the railroad tracks next to the fairgrounds to the Globe Hotel for breakfast. By ten o'clock, Mom and Mary arrived.

After checking in with me, they would go to the girls' 4-H building. At noon, others arrived with a picnic lunch in the trunk of their cars. Nowadays we call it tailgating. Sometimes it was just Mom, Mary, and me. Other times the Van Nices and the Kempers joined us. On show day, Uncle Jim and the whole Shepard family would be there.

The morning of show day, I was excited and woke up early. I walked with Mark and several others to the hotel. The show started at eight o'clock. The first class was for the Angus breed, then shorthorn, and finally Hereford.

They started with the lighter weight animals. Curly was a mid-weight and Whitey a heavy weight, so I was able to show both of my calves. As I mentioned, my animals were born and bred in South Dakota. They needed to be long, tall, and lanky range cattle. I placed near the bottom of the class with Curly and the middle of the class with Whitey. They were the type of animals which, thirty years later, probably would have won the show.

The boys who had fathers in the cattle business had an advantage. My equipment was sufficient, but not as fancy as that of my competitors. Their black and white leather halters and aluminum show sticks out-classed me. I didn't even have a cowboy hat!

When the show was over, the last day of the fair was the cattle sale. For the last time, I washed and curried my calves. After leading my calves through the sale ring, I took them into a waiting truck for the ride to the West Liberty sale barn. My calves were weighed and marked, the halters removed, and driven to a pen with the other calves. Some calves were purchased by families for meat, some were purchased by companies for advertisement and support of the 4-Hers. Most ended up being resold at the next auction and purchased by a packing company.

Some people asked if I felt sad about selling my animals. I told them, "No." I never became so attached to my projects to acquire that feeling. This was a way of making money. The next week I received my first check. After paying for the feed, I made two hundred dollars. Wow! That was a lot of money.

I was free of chores until October. Uncle Jim bought feeder calves in the fall, and as soon as he received his calves, we spent an hour or so on a Sunday afternoon picking out next year's project. I sat on the wooden fence and watched Uncle Jim walk the calves around the feedlot while Uncle Vernon and Grandpa Shepard evaluated each animal. It took a while before they decided on three or four. I got to pick from among these in the smaller group for next year's 4-H project. Uncle Jim chose good calves, more suited for the judge's eye.

My years showing beef animals continued through high school. Each year I started with optimism and ended with an okay. I never placed higher than seventh in any of my classes.

One day after the show at West Liberty, I was talking with some of my friends.

I made the mistake of saying, "I'll never be as good as you guys because my calves come from my uncle's feed lot. His calves are good but not great."

I turned and behind me was Uncle Jim. I backpedaled as fast as I could by saying, "But his calves always made me money."

It was too late. In the fall Uncle Jim made it his mission to prove to me that buying the expensive calves at an auction, especially those sales focusing on the 4-H show, were not always the best money makers. We went to a calf sale in Mechanicsville, Iowa. All the top breeders had animals there. The prices for the calves were much higher than Uncle Jim's Angus, but I had some money, and this was my last year in 4-H. I bought a steer for $210. Uncle Jim's steers usually cost around $120.

Guess what? The high-priced steer did no better in the show ring than the common feeder animals from Uncle Jim, and he cost a whole lot more. The breeders offered good calves for sales, but they saved the best animals for their sons and daughters or themselves.

I was embarrassed by my original statement and Uncle Jim showed me the high-priced animal doesn't always make the most money. It was a lesson learned and not forgotten. My Uncle Jim had much more patience with me than Uncle V.

My Uncle Jim had a dry sense of humor. He always checked in with me before he left for home from the fairgrounds. One day we were standing in the doorway of the cattle barn. A very large lady, who happened to be his sister-in-law, stopped to visit with Uncle Jim. I knew her and turned to talk to the guys behind me. The large lady didn't visit long and continued her journey through the barn. Uncle Jim and I watched her waddle away.

He said to me, "If you had a calf with a behind like hers, you win the show."

He smiled at me as I tried to stifle a laugh but couldn't. Even some of my friends heard the comment and joined the fun. I'll never forget my Uncle Jim. He was my pseudo-father for years to come.

*

Montpelier Busy Boys 4-H club was small, perhaps the smallest club in the county. Most of the time it included the three Arp brothers, three Hearst brothers, two Van Nice brothers, Dennis McElroy, and me. None of us were wealthy families. The ground we farmed was not the best.

Other 4-H clubs had twenty to thirty members. They had fathers who were as competitive as their sons. Many were breeders and their son's projects also sold breeding stock from their farm. These dads were just as involved with the show as their sons were. This is not saying our fathers weren't involved, but our funds were not as plentiful, and we managed our own projects. Having the fewest animals at the fair did give our club one advantage, though. The fair offered a prize for the cleanest and best displayed projects. Our club won once and placed second twice.

My first year of 4-H, I played my first ever basketball game. I was on the junior team. Our first and only game was in March. I was eleven. Although I had watched many games at Muscatine High School, I never was on a basketball floor. Mom and I arrived at the high school to play our game but were shuttled to the grade school gym instead. Our coach for the day was Jim Henke, one of the older members. He tried to show us some plays on the side, but there was no room.

He said, "Follow me." The team followed Jim out of the grade school gym through some halls of the attached high school, up some stairs and into the glorious Muscatine High School gym. We climbed down the bleachers to the floor. I thought I was in basketball heaven. We practiced on the hallowed high school floor.

The game that followed, though, was a disaster. I had

watched high schoolers pass and shoot, but when I got on the floor, I was scared to death. The ball felt too big for my hands. When I caught the ball, I panicked. Before no one guarded me, I took two dribbles and stopped. Now what? Instead of throwing it to someone else, I took a step. The whistle blew. Our poor team of little guys lost 8 to 4.

The year before I joined 4-H. Our senior team fared much better. All the high school's games were over and high school players were eligible to play 4-H ball. The tournament was held in the high school gym and not the one we played in. The tournament included players from Wilton, Durant, Muscatine. and West Liberty.

Montpelier 4-H was at its athletic peak. We had three Henke brothers and three Schroeder brothers. They all had basketball savvy. In fact, Clair and Jerome Henke played for the Muscatine Muskies High School team, as did Greg Schroeder. Greg went on to play for the Iowa Hawkeyes. Montpelier won the tournament handily for two years.

The next year Clair graduated and the Schroeders moved to town. Montpelier quickly returned to the bottom of the tourney. It was fun while it lasted. It proved our small club could win if we had the right players. Our problem seldom did we have the personnel. I was never very good at sports, but I found out later my contacts through sports were a benefit to me.

When it came to softball games, we were mediocre. It was fast-pitch softball, so pitching comprised most of the game. If the team didn't have a pitcher, it was difficult to win. I tried; I could get the ball over the plate, but with no velocity. In the later years, Jack's brother Joe was very good, and we won a few games on the strength of his arm.

There were two other events the club had each year. One was for new members and newly elected officers of the club. It was held in October, and the highlight was a hot dog roast and apple cider feast. Nothing spectacular, except the apple juice or cider was pressed by the members.

On the Sunday afternoon before, we'd go to someone's orchard and pick apples. One of our leaders, Lyall Paul, owned an apple press. In about two hours we had the apples washed and ready for grinding and pressing. It was fun. I look back on the experience and wonder, why didn't somebody get sick? When we gathered the apples, we picked up fallen apples, apples with worm damage, some might have a rotten spot, and good apples from the tree. They were all mixed together for the juice. I guess Mrs. Paul did heat the juice to supposedly kill whatever bad bugs there were, but anyway, it tasted good.

The other event was held the last Sunday in June. It was called the club tour. All the families met at Wild Cat Den State Park for a picnic lunch. Afterwards, we formed a caravan and visited each member's home. Each 4-H member displayed his project and explained its care. Some members' homes, I noticed, were unkempt. We all had barns and livestock in small pens. I realized my fellow members were no better off than me.

One of the bright spots of my 4-H career was when Jack Van Nice and I participated in the 4-H demonstration contest. This wasn't a livestock show, but a contest in speaking and presenting an idea. Mr. Van Nice, Jack's dad, thought our club should try the event. Until now the boys club had never participated. The girls club did every year. The problem was, what would be our subject?

At this time the swine industry was changing their image from a market hog covered with layers of fat, which was used for lard, to a much leaner animal. Vegetable oils were taking the place of lard for baking and cooking. The extension was trying to educate the farmers to change their breeding practices. It was the top subject.

Jack and I decided to tackle the idea. The process was to take a measurement of backfat on a hog before he or she was slaughtered. The extension office provided the material. We had to provide the speech.

One of the problems was there was no way we could work on a live hog for the demonstration. We needed a model. Fortunately, a neighbor, Mrs. Helen Stroh, was very artistic. She and her husband built a lifelike hog out of wood, chicken wire, wet paper towels, and plaster of Paris. She painted it like a Hampshire hog. With our model built we practiced and practiced, honing our speaking skills.

We presented our demonstration, called "Probing for Profit," at the West Liberty Fair. There were four other clubs with teams: Sweetland, Bloomington, Fulton, and Montpelier. Fulton had won the previous four or five contests and represented Muscatine County at the State Fair. They were confident. Their subject was electricity. They had little buildings built and bulletin boards to display their subject.

Jack and I were the first team to present. We explained our subject. We answered the judge's questions afterwards. Now all we could do was watch the other teams and hope. Bloomington's team was good. Sweetland's team forgot their lines, and Fulton was perfect in their presentation. When Fulton was questioned, they faltered.

The contest finished. We waited for the judge's decision. He started out with comments about each team. When he talked about Fulton's team, he was very critical. He gave Jack and I good remarks. Finally, he revealed who would be representing Muscatine at the State Fair. In second place was the Bloomington team.

He announced, "The Montpelier Busy Boys team with their subject matter and their excellent knowledge of the subject will be first."

Wow! We had beaten mighty Fulton 4-H. Jack and I were taking our plaster of Paris hog to the Iowa State Fair. Our efforts were rewarded. It was a first for the Montpelier Busy Boys 4-H club.

A week later we presented our demonstration at the fair. We, along with the team from Bloomington, received a blue ribbon. It was such an honor for us and our club.

The Montpelier Busy Boys 4-H club was not the biggest, or the richest, or the most athletic. We were a group of boys who all were proud to be members of the club. We stuck together. There was no rivalry within the club. I will never forget my club and the friendship it provided. I feel my experience in 4-H helped me express myself. When speaking or presiding over meetings, I have seldom been nervous. These experiences taught me to be a leader.

Mom and Me

MY UNCLES THOUGHT their sister should continue her life as if her husband was still alive. Maybe some thought she should find a job in Blue Grass or start teaching again. They didn't realize mom washed the milk equipment every morning. (Milk equipment was just rinsed at the night milking).

Wayne depended on her. We were in a livestock-share agreement, and she thought she should help some. Rural women seldom drove to town for work. In Mom's case, not only did she have a young son, but weather and roads would be a deterrent.

It became Mom's release from her daily problems to become more involved with the Montpelier Farm Bureau Women. She was elected chairperson of the group. Soon she was attending County meetings in Muscatine. The many nights she was away Mary babysat me, sort of. She had homework to do. In high school she was a member of the Thespians, a group who staged plays. Mary worked behind the stage many times. On the days leading up to a production, she didn't come home and would stay at Sue's or with her friend, Nancy Hahn. I spent many evenings alone at home.

I guess I was safe. Our house sat back from the road and was not visible. I didn't dare lock the doors. If I did, Mom couldn't get in when she came home. There were times I thought I saw a shadow move and I heard something outside. I'd grab my Mexican baseball bat and keep it in bed with me until I knew Mom was home.

Although I was in bed, I seldom slept until she arrived. I wouldn't say anything because I was supposed to be sleeping. I didn't dare tell her I was afraid. My mother needed her women's group, and I wasn't about to spoil her need by being a sissy.

To me, she was a brave lady. She had to settle Dad's estate. It was difficult since she knew very little about our finances. Dad took care of the money. Oh, she knew it took money to plant crops and buy feed, but she didn't realize how much. On top of those problems, she had to raise a young boy, counsel a teenage girl, and maintain our 1951 Studebaker. Automobiles were to be maintained by husbands. She knew she should change the oil every 1000 miles, but little else. Dad always took care of the car. He traded cars without her knowing it.

As I look back, I realize I didn't have a clue about her situation. I was still living as usual. I don't believe I caught on to the gravity of her problems until high school. Finances became significantly tighter. I could tell she was spending funds on her children instead of herself. Mary sensed the money problem. Although she was enrolled at a private college on the advice of Mom's friend, Alpha, she surmised it was too expensive. Her second year Mary switched to Iowa State Teachers College where the tuition was much less.

Mom's driving skills were not good. She was a very timid driver. During the winter we would travel highway 22 to Sweetland. Although it was gravel, it was the safest route. Getting to the highway was the trick. The gravel road east of our place turned north at the top of a hill.

One morning, the curve was slick from packed snow

and ice. Mom approached the hill slowly. As we entered the curve, the car slid to the lower side of the curve and would go no further. Mom decided to back down the hill. She slowly backed up. I was sitting in the middle of the front seat and Mary on the right side. All I could see was the field next to us. I knew there was a steep ditch in between.

I was scared. Mary scolded me for being so wimpy. We finally made it to the bottom. Now what? Do we try again or back up the opposite hill and go home? Luckily for us, a neighbor drove up. He rolled down his window and asked if he could help. Mom explained our problem.

"Edna, you got to gun your car on this hill."

Mom still wasn't convinced.

"Would you drive our car up?" Mom asked. "I'm afraid I can't do it."

"Sure."

The neighbor parked his car and I crawled into the back seat of the car. Mom moved over. He backed the car up the opposite hill as far as he could and put the car in drive. He floored it, and zoom, we were to the top of the hill. He parked the car and got out.

"I can walk back. You go on to church."

Mom thanked him. We stayed until we saw the neighbor return safely to his vehicle. He had no problem making the hill. He waved as he drove by.

By afternoon, the ice had melted, and we safely returned home. Mom learned a lesson that day. She became determined to be a more confident driver because she was the only driver now. In a couple of years, Mary got her driver's license and Mom got some relief.

Although the Studebaker was a new car, the gravel

roads were rough on it. The left front fender began to ride low. This was before power steering and Mom had difficulty steering. Mom didn't have a clue about what was wrong. One day she was attending a Farm Bureau meeting and she mentioned her car problems to Helen Stroh. Mom picked her up to go to meetings

"I bet Hank can fix it," Helen said. "I'll send him up when he gets home from work."

The Strohs lived about a mile away. Hank was a machinist and worked in town. When Hank arrived, he raised the hood, and the problem was obvious: The left front shock absorber was detached from its mounting. Hank quickly threaded a nut and washer on the end, and the Studebaker was fixed.

Another time after my piano lesson, Mom and I went to downtown Davenport for some shopping. We parked on the levee lot. As we were walking away, a man stopped, rolled down his window, and said, "Ma'am, there is a stream of gas coming from under your car. I think you should look at it."

We hurried back and sure enough, gasoline was trickling from the gas tank. There was a growing pool under the car. Mom didn't panic; she drove the car to Gramp Studebaker garage on Third Street. Mom knew the manager and he quickly found a spot for our car. We waited a few minutes for him to find the problem.

He said, "The gas line has broken free of the tank. We'll have to remove the tank, drain what gas is left, and install a new line. It will take an hour or so. I asked my mechanic not to leave until it is fixed."

We walked the two or three blocks to Petersen's Department Store, did our shopping, and made it back before the shop closed. It was after five before the car was ready. The manager asked about how we were getting along without Dad. He said anytime Mom had problems, she should contact him.

Mom thanked him for his help. He opened the garage door and we drove away. Mom looked at the gas gauge. It registered empty. We had lost an entire tank of gas. We stopped at a little service station on Third Street. It was after sunset when we got home, and I still had chores to do. It was one of those times Mom really needed Dad.

Later the same summer, Catherine Thoene, our cousin in South Dakota, asked Mary to be an attendant in her wedding. It would be a long drive, but Mom thought we should go. Mom and Catherine's mother were close friends before Dad died. Since Mary wasn't old enough to drive, Mom drove all the way.

Dad could have driven the distance to Bonesteel, South Dakota, in one day, but Mom thought she could not drive that much in a day, so the trip took us two days. It was over six hundred miles. We spent the first night somewhere in western Iowa, across from a county fairground. The noise didn't shut down until after midnight. I sat in front of the motel wishing I could go to the fair; there was nothing to do at the motel. I guess Mom and Mary were reading or talking. Whatever they were doing it didn't include me. Television was not one of the amenities in a motel. They were just little houses with a bed or two and a bathroom.

We arrived at the Thoene Ranch by early afternoon. The ranch house was huge. It had many bedrooms since in earlier days hired help stayed in the ranch home. We were herded to a couple of bedrooms upstairs. Of course,

Mary had the rehearsal to attend. Mom and I were invited to tag along. Again, I was the only young boy there. I'm sure I was bored out of my mind. I remember walking around outside, but I was afraid to venture too far. Bonesteel was a strange town with a super wide paved main street and dirt side streets. It was a typical western town.

The wedding was a big South Dakota bash. It was held in the community center which was a large Quonset building. Mary was one of six or seven attendants. Afterwards there was a big dance and plenty of booze. The Thoenes were a prominent family in the aera. Somehow, after several hours, Mom and I escaped the party and drove out to the ranch, ten miles out of town. We stayed for a couple days. Mom knew many of the adults living near Bonesteel through her travels with Dad and the Thoenes. When she visited those friends arranged potlucks for the meals.

I have no idea what my sister was doing, but Logan, Catherine's brother, tried to entertain me. He was the ranch manager. I rode around with him in his pickup when he checked his cattle.

When we stopped at a machinery dealer in Bonesteel, Logan knew the owner was a fisherman, and he thought I might like to go fishing. Logan asked the owner if he would take us out on the river. The owner replied he was going out the very next morning and we could come along. I figured it would be better than sitting with a bunch of women. Before we left the store, the man at the store told us to catch some frogs.

Logan asked, "How many?"

"As many as you can get."

Just below the ranch house was a little pond.

Logan said, "There must be some frogs there. We'll take a flat board down. When you see a frog, you whap him with the flat side of the board and bury him in the mud. Then you reach under and pick him up."

We waded into the pond, and it became clear we had to go deeper. Logan removed his jeans and his underwear and waded in. I reluctantly followed. I could see the house; there were women in there. This didn't seem to bother Logan. Since I was not very good at catching frogs, Logan did most of the work. We caught several and put them in a jar.

The next morning, Logan woke me up at five. We drove to Bonesteel and met the implement dealer. From there we drove to the reservoir. The Corps of Engineers was building the Fort Randall Dam on the Missouri River, and they were beginning to fill the reservoir. In places it was over a hundred feet deep.

It wasn't the kind of fishing I thought it would be. We climbed into the implement man's boat. Instead of a pole and line, he used trout lines. He'd string a rope from a couple of almost submerged treetops and dangled hooks from the rope. We went from line to line. Logan helped put a hapless frog on each hook. I just took the lid off the jar. I didn't want to touch those slimy things.

That morning there were very few fish on the line. By eight or nine I was very sleepy. Logan and his friend talked. The drone of the outboard motor put me to sleep. When we got to shore, both men kidded me about being a poor fisherman.

Now the same reservoir covers many square miles and I'm sure where we fished that morning is covered by many feet of water. The implement dealer and Logan are no longer with us. All that remains is a memory of a young boy boating through half submerged trees while the sun was just rising, being very bored.

I returned to South Dakota several times over the years. Each time Logan and his family have entertained my family and me like celebrities. Logan was a good rancher and loved to show his spread to others, especially his second cousin from Iowa.

On our second Christmas without Dad, Mom and I picked out a tree from the Y Men's Lot. It was a scraggly Douglas fir. Again, Hank Stroh came to our rescue and built us a tree stand. Mary helped string lights and hang the shiny balls. The toughest part was draping the aluminum icicles. We tried to save some from the previous year, but they were usually a big ball of shiny metal.

I think Mom got the hint I wanted an electric train. I was twelve and I thought I could figure out the set-up by myself. I wouldn't need one of my uncles to help me. I had all the Lionel catalogs from the Parker Department Store and basically wore out the covers on the catalogs by thumbing through them so often. It wouldn't be a problem to know which one was best. I knew Mom couldn't afford the fancy sets, so I just mentioned the smaller, less expensive sets. Lionel offered a train set for boys who were just getting started.

Sure enough, on Christmas morning there was a big orange and blue box with Lionel written on the side. The card taped to the box read "To Bob from Santa." Now I had long ago figured out the Santa ruse and I knew it really came from my mother. Inside the box the set contained an oval loop of track, a steam engine which smoked, a coal tender, an orange box car, a gondola, and a red caboose. It wasn't much, but it was a beginning. I was so excited I didn't want to open the rest of my gifts.

As soon as we returned from Christmas at Grandma Shepard's, I set the track up in my bedroom. Since my bedroom was small, part of the track ran under my bed. I fashioned a tunnel out of a brown paper grocery sack and books, then I lay down and watched the engine puff through the paper tunnel.

From that day on I would save my allowance and any other money I made to purchase a new car or an accessory. I dreamt of a larger engine or even a diesel, but they were $75 to $125 apiece, way out of my budget.

Eventually, I moved my layout to the basement, where I had more room. I had several feet of track with switches. I purchased a 4x8 piece of plywood and painted it green.

On Saturday night after Thanksgiving, Alpha stopped out. I was in the basement when she arrived. Mom sent her down to surprise me. Alpha asked me questions about the train and how it worked. I explained everything in detail as if she didn't have a clue how it worked. She soon discovered how much I treasured my train.

Later in the evening, Alpha asked what I would want for Christmas. I went to my room and dragged out my well-worn 1953 Lionel Catalog. I told her I'd like anything for my train set. I was thinking maybe a boxcar or flat car.

"Now what would you like the most?" she asked.

I flipped a couple of pages and replied, "I would like the automobile car, or the milk car, or maybe another box car."What I really wanted was the self-unloading cattle car, but I also knew it was way too expensive to ask for.

Alpha smiled and asked again, "Bobby, if you could have any accessory for your train, what would it be?"

I turned to the two-page spread showing the cattle car. "I would like this unloading cattle car, but it costs a lot."

Alpha smiled. "I'll see what I can do. Maybe it won't be that costly at some stores."

I never in my wildest dreams thought I would get the self-unloading cattle car. When Alpha arrived for Christmas, I was outside doing chores, so I didn't see her unload her gifts. We went to church as usual. Because Mom had to go to Grandma's early in the morning, we decided to open gifts on Christmas Eve when we came home from church.

Somehow, Alpha had hid her gift to me behind the tree. It was the last package I opened. As I tore the paper, I could see the orange and blue box. I knew it was something for my train. I read the end of the box and it read "Lionel self-unloading cattle car."

I couldn't believe it. Slowly I slipped the whole apparatus out of the box and laid it on the floor. There it was: a livestock car, a loading dock with pens, and eight little black Angus cows. I wanted to take it downstairs right away and hook it up.

"You must wait until tomorrow. It is already eleven o'clock," Mom told me.

The next morning, I quickly did my chores and headed for the basement. Because of the Shepard Christmas, I didn't install my self-unloading cattle car until the day after Christmas. Alpha would have to wait until her next visit to see how the cattle unloading car functioned. I believe Alpha got as much fun out of my excitement as I did.

*

In the little village of Sweetland stood our church, some houses, and a sale barn. The sale barn was a place where farmers brought livestock to be sold to others or to buy new stock. The barn had many pens for keeping each farmer's livestock separate. In the middle was the sale ring. It had rows of benches were the men sat. An auctioneer sat high in front and cried the sale. Crying the sale meant pricing the animals with a chant.

The owner of the barn was a member of our church. He soon discovered he needed a lunchroom for the buyers of the livestock. He proposed to the women of the church, if they served lunch to his workers, buyers, and the farmers who came to the sale, they could keep the profit minus the heating bill and electricity. So began a long relationship between the church ladies and the sale barn.

The lunchroom was a converted hog barn. The women divided into six groups. Each group worked a Saturday at the sale barn. Each group had a different menu. It was a win-win for both parties. The women financed their many mission projects and the owner had food for his men and customers. The food was all prepared by seasoned farm wives. It was always delicious.

Mom was part of group five. On her Saturday to work she'd leave around eight, seven if she was her group's chairperson. Mary and I were left at home. It would be after four before Mom would return.

One January Saturday, she was chairperson. Mary was in Muscatine preparing for the school play. I would be home alone. Mom decided I was too young to stay home alone for that long of a time, so I should go to the sale barn with her and our neighbor, Catherine Van Nice.

It was the typical January thaw day. Friday had been warm for Iowa, in the 30s. The first hour or so was not so bad. I helped carry in supplies and had a doughnut. The workers started to come in for an early lunch, so I was shooed out. I wandered round the barn and the pens until some gruff man told me to get out because it was too dangerous for kids. He was probably right.

I went inside the sale ring and sat. It was warm there, but boring. Around noon some other boys came with their dads. We played for a short time. The sale started at one and most men were in the barn. I returned to the lunchroom. The women were taking a break. Mom gave me a roast beef sandwich, their group's specialty, and a piece of pie. She suggested I go watch the sale.

As I crossed the drive, I noticed the wind picking up from the northwest. There were a few snowflakes in the air. I went inside to the sale ring. It was warm in there. The livestock buyers were seated around the ring with their tally sheets. All the boys were gone, so I was alone. The top row of seats was open, so I climbed up. It didn't take long for my eyes to become droopy. I laid down and took a nap.

I don't know how long it was, but when I woke, the livestock had changed from feeder pigs to cattle. I looked around and no one seemed to care what Bobby was doing. It was very boring.

Around four, I exited out the back stairs and went to the lunchroom. Mom and the other ladies were cleaning counters and packing food away. Just a few stragglers were eating their last piece of pie before going home.

One commented, "I'll bet it's dropped twenty degrees since noon."

Another commented, "I just looked at the thermometer. It's fifteen degrees out there."

Mr. Hetzler, the owner, came in to pay his bill. He told the women he thought they should close and head home. Mr. Van Nice stopped to pick up Catherine. Mom was the last to leave. It was becoming dark when Mom and I went to bring the car around for the final loading. She turned the key.

The Pontiac groaned. It didn't start. She tried again. Nothing. I sat in the car while she walked over to the sale barn office to find help. There were very few men left, and they were busy. The wind was now howling out of the north. Mom was about to call one of the uncles when Cliff TeStrake, a church friend, stopped and asked if there was trouble.

I said, "Yes, Mom is over at the office trying to find help."

"Let me open the hood and see what I can do," he said. "Turn the key." The car groaned again. "The battery is dead. I'll find someone with jumper cables. You tell your mom I'll be back."

In a few minutes, Mom returned. Cliff was right behind her with jumper cables and his pickup. He and another man connected the two batteries. Mom turned the key, and the Pontiac roared to life. Cliff came to her window and said, "Edna, don't turn off the engine until you get home. Your battery will be recharged by then."

Mom replied, "Thank you very much. I don't know why it wouldn't start. It's a new car."

Cliff said, "Your car was pointed into the wind. It froze the battery. I try to never face into the wind. It is supposed to be below zero tonight. If you have a heat lamp at home, I'd place one over the battery to keep it warm."

We made it home. After unloading the car, I still had chores to do. When I finished, I helped Mom locate a cord and heat lamp. Our fingers were very cold before we accomplished our task. The next morning was church. It was a minus four degrees. Mom tried the car early to make sure it would start. It started.

We made sure we never parked the car facing into a cold wind again. We also placed a heat lamp over the battery when the nights were frigid. I remembered the suggestion from Cliff for the rest of my life. It was one of the lessons I couldn't receive from a book.

I don't believe I ever was required to go with Mom to the sale barn again. Her next stint was in March. I convinced her I would be okay at home alone. She left me some lunch and I began my long day. After watching cartoons all morning, I became bored watching the afternoon old movie. Television sportscasting was in its infant stages. Networks didn't realize the value of sports.

By three, I was playing a solo game of Monopoly or checkers. I had tired of my train in the basement. It was too early to start chores. I dug around in the top drawer of my clothes chest, where I kept all my treasures. I located a magnifying glass, a present from Alpha. I had read if you focused the lens just right, you could start a fire.

The sun was out, so I went to the living room where the sun was the brightest. I played around looking through the lens at different items. I stared at the fiber in the rug. I pulled the glass up until it made a bright spot on the rug. Interesting! I focused a little more. The sunlight made a bead on the rug. Suddenly, the rug started to smoke. I quickly pulled the magnifying glass away and snuffed the blackened area with my shoe.

It was about as big as pencil eraser. Now what would I do? Mom would be mad if she found out. I ran to the kitchen, found a sponge, and rubbed the blackened spot until it was gone. The air smelled of burnt wool, so I found some room freshener in the bathroom, and sprayed all around.

I had only one problem left. The hole in the rug. I brushed the fibers up around the depression. Luckily, the carpet was sculptured. Soon I had everything back in order. Mom would never know, at least, not that night. That evening I glanced many times at the burn, but I couldn't detect any sign of it.

A couple of days later, when I came home from school, Mom was vacuuming the carpet. She said nothing. My patch had withstood the test. I never told Mom of my experiment, but I will confirm you can start a fire with a magnifying glass. Sometimes you just get lucky.

It would be another six weeks before Mom had to work the sale barn lunchroom. By then it was early May, and I could be outside, probably driving tractor.

Ron

ONE EVENING IN LATE April or May, a green Plymouth sedan drove up our lane. I was ushered to my room. Mary didn't want me to embarrass her. I watched from my bedroom window. The young man walked to the door.

Gosh O Mighty, it was Ron Havemann, and the guy in the car was Don Huff. Both were stars of the high school basketball team. Don was driving and he had a girl almost sitting on his lap. What were they doing at my house?

I could hear Mom and Ron talking. Next, I heard Mary coming down the hall. I tried to listen through the wall. I heard Mary say something about her little brother. Suddenly, Mary knocked on my door.

"Ron wants to meet you."

I followed her to the living room. There as big as life was Ron Havemann. He seemed taller than on the basketball floor. He was wearing his purple and gold letterman's jacket with the big gold M on the side. I shook his hand.

He said, "I have a little brother, too. Someday I'll bring him along."

I replied, "Okay."

That was all I could say. I was star struck. To me he was as good as a movie star.

During the summer Ron visited Mary many times. Because he was on the Muscatine high school baseball team, he'd play catch or hit balls with me. Many times, the Kraft kids joined us. Ron loved to show off his ability to catch balls behind his back or throw them under his leg. He was my sports idol.

The next season, the Muskies basketball team started off strong. They were undefeated through January. The town was excited. Their biggest challenge was the Davenport High Blue Devils. Davenport was the largest school in the state, and they had won the last three state championships. Muscatine had to play them in Davenport's gym. The night before the game Uncle Jim called. He had tickets for the game.

We entered the huge gym. The adults had reserved seats. Mary went with her friends to the visitor's section. Cousin Jane and I were ushered over to the grade school section with all the Davenport children. We were given strict orders not to leave our seats until after the game.

Unfortunately, it was the Muskies' worst game. Davenport controlled the game. The Muskies lost.

The loss would be a blessing in disguise. The Muscatine team started on a drive to the state tournament. In 1954, there was just one state champion. All schools, large or small, competed for the title. Muscatine quickly ran through the small school teams and made the state tournament.

The state tournament was held at the big field house in Iowa City where the Iowa Hawkeyes of the University of Iowa played. Mom and I listened on KWPC-FM. The Muskies won their first and second games.

Uncle Jim called early in the morning. He had acquired two extra tickets for the game against Council Bluffs. He asked if Mom and I would like to go.

"Like to go. Are you kidding? I'd love to go."

I had a day to get ready. My plan was to throw confetti

if the team won. I tore up any paper I could find during recess at school. By game time, my grocery bag was half full. We rode to the game with Uncle V. I was in awe of the size of the field house. I carried my bag of confetti. Uncle V probably thought it was stupid, but he didn't say a word. I'm sure he hoped I'd lose it somewhere.

Muscatine never trailed. Toward the end, I started to toss my confetti in the air every time the team made a basket. I emptied the bag at the last buzzer. The poor people in front of me received my paper shower. Since they were all Muskie fans, no one was upset. Muscatine was going to the finals against Des Moines Roosevelt.

The finals were Saturday night. Mary and the uncles attended. Mom and I were going to listen on KWPC.

Alpha came from Indiana to be with us. The local news announced the game would be broadcast for the very first time on TV. The local stations decided not to carry the game, but a Cedar Rapids station would.

After I finished my chores, Mom and I took a long stick and turned the TV antennae towards the north. The station came on the screen, but it was very snowy and sometimes faded in and out. I sat as near as Mom would allow.

The game was close. At one point, Mary's boyfriend stole the ball three times and scored six points. Muscatine was ahead by eight.

Alpha said, "The game is over. They will win."

She was right. The Muscatine Muskies were State Champs. The celebration began. Mary came home the next day so hoarse she couldn't talk.

The victory assembly on Monday was broadcast on the radio. Mom let me listen when I got home from school.

Although I was the kid brother, I was as excited as Mary. By this time, I knew most of her girlfriends and, of course, Ron.

Mary was a senior, and college was her next goal. Alpha suggested Cornell College in Mount Vernon, Iowa. Mary started there in the fall. It was a big event for my sister to attend college. When Mom and I drove her to Cornell, we met her roommate from Cedar Rapids. Although my sister and I did little together, she was my big sister. I was enthralled by the school and all the people from all over the country.

One Saturday in October, Ron stopped out at our place around ten. He asked if I would like to go to Cornell and see Mary. It was Cornell's homecoming. Mom quickly fed me lunch and off we went. It was about an hour drive to Mount Vernon. Ron bought the tickets, and we sat in the visitor's section.

Mary had no idea we were there until after the game. Ron and I met her as she walked back to her dorm room. I wasn't dressed warmly enough and shivered most of the time.

Fortunately, the Cornell Rams won. There would be no classes on Monday. Since her roommate was from Cedar Rapids and going home, Mary decided to come home also. Mom could drive her back on Monday.

After a year at Cornell College, Mary realized the tuition was too expensive for Mom. She transferred to the Iowa State Teachers College (now the University of Northern Iowa) in Cedar Falls. It was a state school and fit Mom's budget.

Mary started right away and attended summer classes. She came home in late July for a month before classes started again in September. Her being home was more of a guest appearance than visit. I had to depend on my peers to teach me the ins and outs of high school.

High School

MOM DID THE BEST she could to guide a budding teenager. She made sure I attended church each week. My days at Hazel Dell were about finished. Mary was away at college. I was green as a stalk of corn in July when I started high school. My peers were my guidance for my social life. I didn't feel Mom was up to date with the modern world when it came to girls other than my cousins.

I had a great time in high school. My activities at MHS were FFA and mixed chorus. In FFA I achieved the degree of Iowa Farmer and sang in the FFA State Chorus. I received a state honor for farm mechanics and electricals. My instructor, Mr. Bietz, knew I had accomplished many repairs for our farm and had me apply for the state award. He realized I didn't have the fatherly advice many of his other students had. I was honored at the state convention for my skills.

My choral experience in high school was minimal. I was accepted because I could carry a tune and was a tenor. In my junior year, I did make All-State Chorus. It was held in the KRNT Theatre in Des Moines.

I didn't realize the significance of the event until many years later. I found out there were hundreds of students throughout the state who tried out for the honor chorus. I just happened to be fortunate to be one of those students. Our chorus staged a musical every spring. By the time I was a senior, I was good enough to be given the small part of Omar in the musical *Kismet*.

My grades were decent, mostly Bs with a few As scattered around. I think it was because of my advantage, or disadvantage, of being isolated. I came directly home every day after school. By the time I finished my chores, I had little time to play. I would study or work on homework, watch some TV, and be in bed by nine. Morning came fast since the school bus arrived at 7:20 in the morning and all chores had to be finished.

I was able to drive in November of my sophomore year. I quickly became popular with the girls because of this ability. I dated some, but my popularity soon faded.

Most girls I dated lived in town and they found boyfriends who could date during the week. I lived eighteen miles from town. By the time I was finished with my duties at home, it seemed silly to drive a half an hour to town just to cruise around and maybe hit Leu's Ice Cream or Cole's, then drive thirty minutes home again. My dating was limited to Friday, Saturday, and Sunday afternoon. Even most of my farmer friends lived closer to town.

It wasn't until I took Mrs. Fogarty's senior English Composition when I noticed a tall slender girl with brown hair. She sat in the middle of the room, and I sat toward the front, the problem of having your name start with a B. I don't know when we first spoke to each other, but we must have at some point. Her name was Jane Klatt. At the end of our year, the seniors had a dance where girls asked boys. Jane stopped me after class and asked if I would go with her.

I would have, but at the time I was having a few dates with our foreign exchange student, Josette, and Josette had just asked me to go with her. I had to turn Jane down.

The few dates with Josette were entertaining and fun, but I knew she was a dead end. At the end of the summer, she was returning to Germany. The family Josette lived with was considered upper class in Muscatine. A week after she had returned to Europe, I discovered her American family had thrown a going away party and I wasn't invited. I guess Farmer Bob was not good enough.

I brushed it off as a lesson learned. I wasn't what the snooty wealthy girls and their parents desired. The boys with the fast cars and money were better.

Graduation was over and summer was coming. The YMCA had record hops called Fun Nite in their gym all summer. This was the place to meet your friends. The very next Friday night June 26, 1959, I attended Fun Nite at the Y. The girl I had turned down for the dance at graduation time was there, and I asked Jane for a dance.

Jane's dance moves were A or maybe B+ whereas my steps were a C-. Luckily for me, she tolerated my attempts. During the last dance I asked if I could give her a ride home. She said, "Yes."

We went to the teen hang-out, Cole's. As we walked in the door, I checked my wallet. I had exactly one dollar. It would be enough, I surmised. Hamburgers cost 30 cents, Cokes 15. That would be ninety cents. All right!

When the waitress asked for our order, Jane said, "I'll have a tenderloin and a malt."

Jane claims I answered, "Really!"

A tenderloin cost 35 cents and a malt was 25 cents, which equaled 60 cents. I must have had a shocked look on my face. I did the math and said, "I'll have a Coke."

Jane smiled at me and realized her suitor was short on cash.

She said, "Change mine to a hamburger with pickles, mustard, and a Coke."

I replied, "I'll have the same."

I now could cover my expenses. I was grateful and surprised. I went home that night thinking; this girl is a keeper. I never dated another girl.

On our next date, I drove Jane to a drag strip in Cordova, Illinois. She wore a red and white checked dress with lots of petticoats. We sat and watched dragsters roar down the strip. I'm sure it was boring for Jane. Honestly, it was a little boring for me. It was just a place the guys all talked about, and I had never been there. I never went again.

I asked her to many events after that. She told me years later she wondered if she'd ever get kissed by me. It was at least the fifth or sixth date before I puckered up. She reminds me still of the first time I brought her to the farm. I surprised Mom, but we probably would have had boiled hot dogs anyway. Mom was not a gourmet cook. Jane still reminds me of our supper.

I'm glad Jane thought I still had promise. I think she thought she'd stick with me until fall since she was going to a secretarial college in Milwaukee. I could be disposed of. She didn't realize how tenacious I was. I never let her go.

Jane and I dated for two years. Mom wasn't surprised when I told her I was going to give Jane a diamond. I saw no reason to delay the obvious. Mom was disappointed we wouldn't wait until I graduated from college to get married, but Jane and I never regretted our decision.

Marriage and Beyond

JANE AND I SPENT two years in student housing at Iowa State. She worked at the student placement office on campus while I attended classes. I never worked part-time anywhere because every other weekend we drove home so I could work at the farm.

Most of the families in student housing were as poor as we were, but it was the best time we had in our young marriage because we had to work out our problems together. There were no parents to bail us out. Our first son, Blaine, was born a month before I graduated.

While we were at Iowa State, we had the opportunity to purchase a farm just south of the home place. It had been an estate and the owners were anxious to sell. The previous owners let us start to remodel the house before we closed on the farm. The house had electricity but no running water. Jane's parents spent hours working to make the house livable.

I graduated in late February. We borrowed a friend's stock truck and moved home the day of graduation. I remember driving down US 30 with Jane nursing Blaine on the way. The three of us lived in the downstairs three rooms. We spent the next thirty-seven years in the house, raising three sons, hogs, sheep, and many dogs.

Jane has always been at my side. She's weathered the ups and downs of farming. She always supported my decisions, whether they were good or bad. The stories we could tell of our sixty plus years together will fill another book, maybe two books. She is and will be my only love.

Mom continued to live at the home place until her health failed. One evening she was to come to our home for supper with Jane's mom and other friends. About an hour before, Mom called and told me she had fallen. I rushed to her house and found her sitting on the edge of the bed.

"I think I broke something," she said.

I took her to the emergency room. They confirmed she'd broken her collarbone. She'd need some extra care. Neither Jane nor I had the time to care for her properly.

Because of Mom's connections through Farm Bureau, we knew a lady who was the manager of a retirement home. She found a room for Mom where she would be able to rehabilitate for six weeks.

Then I had to make a very difficult decision. My sister lived in California. I was Mom's caretaker. Mom needed care I couldn't provide. We were fortunate and found a place in Durant with assisted living. She could have her own room and they provided meals. We moved her in. Later, I would bring Mom home for brief visits. I'd drive her around the farm and the buildings but decided not to let her inside her house. I figured it would be too difficult for her to leave again.

One day the retirement home called to tell us they had transported Mom to the hospital. The next few weeks she was in and out of the hospital a couple of times. I surprised her by picking her up on Christmas Day and bringing her out to our home for Christmas. All our family was there. Our house had several steps, but we had three strong sons and they lifted Mom and her wheelchair into the house.

It was her last Christmas.

In mid-January Mom was back in the hospital. She went into hospice care. We visited her after church and found her sitting in a chair. I will never forget the forlorn look in her eyes. She kept asking me to help her. I was at a loss because I didn't know what to do. A nurse came in and tried to feed her some ice cream, but she couldn't swallow.

It was difficult for Jane or me to stay at the hospital with her. We had no hired help. Our sons were married and lived miles away. We had many animals to care for. We had to trust the hospital staff. It was four in the morning when we received the dreaded call.

"Your mother is dying. We suggest you come immediately," the nurse said.

We hustled around. I called Mary, who now lived in California. She said she would be there as soon as possible.

When we arrived at the hospital, we hurried to Mom's room. It was full doctors and nurses.

I asked, "What are you doing?"

The physician answered, "Her kidneys are failing. We are trying to get more fluid into her."

I was shocked. I knew Mom did not want to die like this. She had had a full life. I called Mary from the nurse's station. Luckily, she was still home to answer my call. She agreed with me to let Mom go. It was Mom's wish. I don't think the doctor was pleased with me, but he and his staff departed.

Mom hung on all day. Several times she would draw a deep breath and nothing. We would just be ready to call the nurse and she would breathe again. I called a neighbor to do our chores. The hogs would survive.

At six o'clock, one of the nurses came to check Mom's vitals. I asked if we should get something to eat. She suggested we hurry because the cafeteria closed at seven. We were just about to step on the elevator when the nurse dashed to get us.

"I believe Edna is taking her last breath."

All day we had seen several of her taking last breaths, but she would revive. To me, Mom was gone. It was only the clinical death left to happen.

"We don't have to be there," I answered.

By the time we returned, Mom was gone. Mary arrived an hour later. She told me not to be sad. I had done the best for Mom any son could have done.

Yes, I was sad to see my mother die. She and I had been through a lot, but I hated to see her suffer. God had called her home. There was no doubt in my mind Mom would be in heaven.

I wondered later, should I have returned to see her take her last breath? Maybe. I have read obituaries how family was at the bedside. It has haunted me for a long time.

There were many friends at her visitation. I talked myself hoarse. As we were leaving for the last time, I was alone with Mom. I stood at the end of her casket and said my last goodbye.

Mom and I had been through many events in our lives. Although I was an adult, I hated to think she was no longer here to talk with. Her last years were not pleasant for her because of her health. Now she no longer was in pain or doubt.

We had Mom's funeral at her precious Sweetland Methodist church. I knew I had to sing her favorite hymn "In the Garden." The organist had to lower it a key because of my hoarseness. I made it through the first verse, then my voice cracked. The organist didn't stop; she just slowed until I gathered myself and finished. We buried Mom next to Dad in the Blue Grass Cemetery.

The end of Mom's life is the most difficult to tell. I'm glad I'm almost to the end of my story.

Jane and I decided to move into Mom's home since it was a ranch style home and on one floor. The Bancks farm became my responsibility. I will live here until I cannot function anymore.

In 2018, the farm received the Heritage Farm Award for being in our family for 150 years. I have spent all my life living and working the ground. I have always believed farmers do not actually own their land. God owns it. It is our task and privilege to take good care of His property. I hope He approves of my stay here.

About the Author

Bob Bancks dreamed up many stories in his hours driving tractor on his family's farm, and after he retired from farming, he had the time to write them down. He is the author of several novels and stories which capture the flavor of rural life. His books are for sale online and in many places around Blue Grass, Iowa, where he lives with his wife, Jane.

www.ingramcontent.com/pod-product-compliance
Lightning Source LLC
Chambersburg PA
CBHW030033100526
44590CB00011B/178